KISS THE GROUND

ALSO BY JOSH TICKELL

Biodiesel America

From the Fryer to the Fuel Tank

KISS THE GROUND

HOW THE FOOD YOU EAT CAN REVERSE CLIMATE CHANGE, HEAL YOUR BODY & ULTIMATELY SAVE OUR WORLD

JOSH TICKELL

FOREWORD BY TERRY TAMMINEN,
CEO OF THE LEONARDO DICAPRIO FOUNDATION AND
FORMER SECRETARY OF THE CALIFORNIA EPA

ENLIVEN BOOKS

ATRIA

NEW YORK LONDON TORONTO SYDNEY NEW DELHI

ENLIVEN™
ATRIA
An Imprint of Simon & Schuster, Inc.
1230 Avenue of the Americas
New York, NY 10020

First Enliven Books/Atria Paperback edition September 2018

Interior design by Kyoko Watanabe

Manufactured in the United States of America

10 9 8 7 6 5 4 3 2

The Library of Congress has cataloged the hardcover edition as follows:

Names: Tickell, Joshua, author.
Title: Kiss the ground : a food revolutionary's guide to reversing climate change / Josh Tickell ; foreword by John Mackey, Founder and CEO of Whole Foods.
Description: First Enliven Books hardcover edition. | New York : Enliven Books/Atria, 2017. | Includes bibliographical references.
Identifiers: LCCN 2017032347 (print) | LCCN 2017041054 (ebook)
Subjects: LCSH: Organic farming. | Climate change mitigation. | Agricultural ecology. | BISAC: NATURE / Environmental Conservation & Protection. | SCIENCE / Environmental Science.
Classification: LCC S605.5 (ebook) | LCC S605.5 .T58 2017 (print) | DDC 631.5/84—dc23
LC record available at https://lccn.loc.gov/2017032347

ISBN 978-1-5011-7025-6
ISBN 978-1-5011-7026-3 (pbk)
ISBN 978-1-5011-7027-0 (ebook)

The burning sand will become a pool,
the thirsty ground bubbling springs.
In the haunts where jackals once lay,
grass and reeds and papyrus will grow.

—ISAIAH 35:7
(NEW INTERNATIONAL VERSION)

CONTENTS

FOREWORD

In the fall of 2003, newly elected California governor Arnold Schwarzenegger appointed me secretary of the state's Environmental Protection Agency. Perhaps the greatest challenge I faced in that job was responding to his order to draft a climate change action plan for California, at that time the sixth largest economy on the planet.

We rolled up our sleeves and developed the Million Solar Roofs Initiative to push the state toward goals of being substantially powered by clean, renewable energy by 2020; new energy-efficiency standards for appliances and buildings to save money and cut pollution from power plants; greenhouse-gas limits and low carbon fuel standards to make cars cleaner and more efficient; a "Hydrogen Highway" program to bring clean hydrogen-powered cars to the state alongside electric vehicles; and a host of other measures, including a regional carbon emissions trading market, in our Global Warming Solutions Act of 2006.

Over a decade later, our efforts have made California a world leader in solving climate change and in proving that we can do so while building a stronger economy. But the nagging fear was that, even taken all together, we would still not be doing enough to help the world avoid the "tipping point" that will occur if the increase in our average global temperature exceeds two degrees Centigrade over historical averages.

If only I'd had Josh Tickell on my team in those days! In *Kiss the Ground*, Josh fills that gap by harnessing the most important sector of our economy—indeed, the source of all human life itself—to address climate change and the future sustainability of a population expected to reach ten billion people by 2050: our food supply.

We should kiss the ground, instead of poisoning and pummeling the very life out if it as modern industrial agriculture does today, because the "regenerative" farming that Josh describes in this book is the only practical way to take enough greenhouse gases out of the atmosphere—not just reduce the continuing levels of pollution we add with every trip to the grocery store or flip of the light switch—before it's too late.

If Josh had been serving in our administration, we would have had him use this book to teach a master class not only in sustainable food supply but also in relearning the lessons that allowed humans to peacefully coexist with every other living thing on earth for so many millennia before the advent of the Industrial Revolution. The changes we were smart enough to imagine and invent have certainly helped human civilization to thrive, to cure disease, and to make a great peanut-butter-and-jelly sandwich, but we were foolish enough to think there were no limits to the use of our fossil-fueled machines and farming methods.

Now we know better, and thanks to books like this one, we know how to evolve yet again with a new kind of revolution that will make us healthier while we make our planet healthier, too. And that may be the most important theme of this book: optimism. We've pushed the planet to a breaking point previously imagined only in science fiction. It is science fact, however, that will help us climb out of the hole we have dug for ourselves, if we demand of each other—and our leaders—that we heed the science of the earth, not only the economics of a very outdated way of doing things.

Shakespeare said, "Nature's bequest gives nothing, but doth lend . . . then how, when Nature calls thee to be gone, what acceptable audit canst thou leave?" If we heed the commonsense call to action in *Kiss the Ground,* our generation will leave a legacy that would have inspired even William Shakespeare to his most laudatory poetry.

TERRY TAMMINEN, CEO, LEONARDO DICAPRIO FOUNDATION,
FORMER SECRETARY OF THE CALIFORNIA EPA

AUTHOR'S NOTE

Working as film directors, my wife, Rebecca, and I, along with the help of many great people, have turned this book into a documentary with the same title: *Kiss the Ground*. While this book provides a breadth of information that could never be covered in a movie, watching *Kiss the Ground* is a deeply emotional experience. If you like this book, you'll love the movie and vice versa.

Both the book and the film are due to a single catalyst. In this case, the spark that helped ignite this fire is a young man named Ryland Engelhart. Ryland and his family run a chain of organic vegan restaurants called Café Gratitude. His father, Matthew, and stepmother, Terces, also run a farm called Be Love Farm, both of which are profiled in chapter 9.

It was Ryland who first began talking my ear off about the power of soil to pull carbon dioxide from the atmosphere and the power of our food to change how we treat our soils. Since that time, he and his wife, Sarah, have moved into the house where my wife and I once lived. With a group of dedicated people, Ryland operates the nonprofit organization Kiss the Ground.

Their mission is "to inspire and advocate for the restoration of soil worldwide." Their activities span the gamut, from producing viral educational videos to creating community gardens, to helping to inform community, state, and national leaders about the power of healthy soil to simultaneously sequester carbon dioxide, store water, and grow more and healthier food. Their mission is bold and critically important for the future of our species. Hence, I am donating 50 percent of the proceeds from this book to Kiss the Ground in perpetuity.

INTRODUCTION

It is almost spring in California.

I part the soil gently. I take my daughter's hand and place the seed in her palm. Together we drop it into the small hole. I show her how to cover the seed. I pick up the watering can and tell her, just as the piglet in the book *What Does a Seed Need?* has shown her, that a seed needs water. Like a bee, my pregnant wife buzzes around us, picking weeds and tending to the new plants.

This is the first garden my little family is planting together. My wife, Rebecca, and I have great hopes for these seeds: clover, sunflowers, corn, beans, peppers, herbs, berries, grapes, fruits, vegetables, trees, perennials, and annuals. Some of these plants will flower and fruit this summer. Others will take years. Still others may not work in our particular soil. After all, it has only been three and a half years since the last spray of heavy toxic chemicals on this land.

We are new stewards. Those who came before us took this small five-acre plot nestled against the foothills of the Los Padres National Forest and drove it as hard as they could with millions of acre-feet of water per year and truckloads of liquids like 2,4-D and paraquat, many of which are the by-products of oil refineries in my home state of Louisiana.

Walking the land for the first time I saw dry, cracked earth, no green between the rows of avocado trees, no insects or creatures of any kind. It was man-made and terraformed. Scorched. A largely lifeless stage upon which to play out a family-sized version of the agricultural cycle that may now threaten the foundation of our species.

We have had little time during the past forty-two months to dig deep into the work of healing this land. Our homestead is over sixty years old. Pipes break, electrical lines are faulty, and gray water and sewage are constant concerns. Trees fall—sometimes on cars. This is a tiny farm but each day brings unique challenges.

Then there's life.

Our first daughter, Athena, was born here on this land just eighteen months after we arrived. She is now two. As I try to show her not to run and jump on the seedlings, I wonder if she will remember any of this.

Here's the thing about planting a garden—you can nurture it but you cannot know what will happen. In spite of the modern mechanized food system we have developed to supposedly create food security for ourselves, nature is still in charge. Our nutritional future, therefore, is only predictable within a margin of error.

My first experience working on farms was through an organization called World Wide Opportunities on Organic Farms (WWOOF). That was over two decades ago. Since then, I have been involved in production agriculture in every major growing region and most states in America. I have worked with some of the largest agricultural consortiums in the world, which represent hundreds of millions of acres of conventional monocropped corn and soy.

But no matter where I went, or what type of conventional agriculture I encountered, there were the same telltale signs of crop blight; brittle, eroding soil; and people suffering from exposure to toxic chemical sprays. In spite of the mind-set of modern agriculture, I continue to hold farmers themselves in the highest regard. Indeed, theirs is not a profession, but rather a calling that requires a mixture of faith and mettle that few possess.

All of this is why this current act of backyard cultivation, this partnership between my family and the nature within which we live, is by far my most political, most radical, and most powerful act as a human being. We are creating a contract, an agreement. It is unspoken but implicit and understood by all parties.

Our part of the bargain is to inspire the soil to live. We must keep it covered with green clover. We must not kill it with chemical insecticides and fungicides, even in the face of infestations. We must become the co-stewards of the billions of unseen life-forms (things with names like mycorrhizal fungi) that give it vigor. And even in periods of drought, we must provide water. It's a difficult deal to uphold.

On the flip side, the soil, and nature herself, has a hefty load to bear. She must take these seeds, seeds that are so small as to seem insignificant, and turn them into mountains of food. Her work is indeed miraculous in the extreme.

There are so many variables, so many opportunities for either party to fail in this contract that I am nervous. I look to the sky. We have been promised the great rains of El Niño. But they have been slow in coming. The winds, however, have been strong enough to rip a number of our large trees up from their roots and deposit them unceremoniously on the ground. Just as a small reminder of who's really running this show.

For the next several weekends, we take a day (or at least a few hours) to sow seeds, check irrigation lines, spread mulch, add compost and hope.

We have great hope.

We get lucky. The rain comes. At first in a trickle and then a torrential deluge the likes of which Noah awaited. Our cracked, dry patch of earth has water. My daughter finally has an excuse to wear her dinosaur rain jacket and her rain boots, rarities in the dry heat of Southern California. She splashes in the puddles. She says that she has seen dinosaurs. A baby one, a daddy one, and a mommy one. She reminds us of all that is good in the world.

I leave home to seek more answers to the burning questions in the pages that follow. It is after a long day on the road that my wife phones to tell me she took Athena to the toy store in our small town. I groan—more plastic, more branding, more garbage.

But then she tells me something surprising. She let my daughter pick two toys, one for a friend whose birthday it is and one for herself. Athena looked at everything in the store. For the birthday boy, she picked a set of gardening tools. For herself, she chose a small green watering can.

Indeed, we have great hope for our little garden.

AS OLD AS . . .

Nothing in this book is new. It is a story that has been told a thousand times throughout every culture. It is the story of the rise and fall of civilization. It involves the philosophical, political, and social oppression of a people through food. It involves the subjugation of nature. It involves humans forgetting their origin and then remembering. It is the story of falling asleep and then of waking up. My hope is this story, and others like it, may be just in time.

Ultimately this is a story about the power of our food. It's not so much about how food affects our health, although the health our food provides is connected to everything herein. After all, one cannot truly have healthy food without healthy soil. It's not an absolute formula for what to eat and what not to eat, although this book provides a critical lens through which to make food choices that could literally save our future.

This is also not a book about climate change, at least not in the way it is being argued about today. This book takes the perspective that carbon is the most basic and essential building block of life. Where there is carbon in the soil, there is water. And where one finds carbon and water in soil, one finds food. Thus this book deals with atmospheric carbon not as a tired political problem but rather as an ecological opportunity that we can access through our food.

At its core this book is about food's ability to alter the world. The premise is simple: Our choice of foods will make or break our civilization. This daily choice holds more power than almost anything we do. Food is power. It is so powerful, in fact, that food is likely at the center of what caused the downfall of the previous forty or so significant civilizations that have come before ours.

One could say this is a book about food, agriculture, soil, climate, and planet. But that would be missing the point. This is really a book about the whole system in which all these things interact. Each of the aforementioned acts as one of many interconnected cycles within this system. For most of us, the access we have to altering those cycles is through what we eat.

This book is also about people. Specifically, it is about a global movement of people who, due to their combined understanding of natural systems, could offer a new framework to feed humanity. In its simplest terms the movement of which I am speaking is called "regenerative agriculture." While it will be better defined in the coming chapters regenerative agriculture essentially means building soil through the four pillars of trees, managed herd grazing and compost, cover crops, and "no-till" agriculture systems. Since building soil requires tremendous quantities of carbon, regenerative agriculture holds the key to moving the carbon that's currently in our atmosphere back into the ground.

A HUNDRED WAYS

I recently had the opportunity to meet and interview a man named Chief Arvol Looking Horse. He is the great-grandson of the famous Native American chief Sitting Bull, who led his people during the resistance to the US Army's "culling" of the Native Americans who first inhabited the place we now call America. Chief Looking Horse serves as a critical link to both our peoples' past. He is also the keeper of the sacred pipe of the Lakota, Dakota, and Nakota tribes, the object that binds them and brings them together in ceremony.

Chief Looking Horse explained to me that his people envision time, nature, and man as a continuous hoop. They call it the "sacred hoop." And they believe it is our responsibility, or rather our job as inhabitants who temporarily walk the earth to maintain, strengthen, and, when necessary, repair that hoop. The bad news, he says, is that we Westerners have broken that hoop by breaking our link to the past, breaking our link to nature, and breaking our links to one another.

Chief Looking Horse wants us to repair those links. In fact, he is particularly concerned with repairing the hoop between ourselves and nature. He says it's a pressing issue. He's even seen it in dreams. It's there that he says the elders have shown him that our mission for this time on Earth is to repair humanity's relationship with nature.

One of the nature-man links we have forgotten is the one that binds what we eat with where it comes from. In the simplest of terms, what we eat comes from the ground. What is interesting is the ground is both a thing and a place. Otherwise known as the soil, this is the place where we stand and build and live. It is also certainly a thing. You can pick it up, throw it, and, if you're a kid, you'll likely eat it.

That instinct to eat soil may have something to do with the fact that, according to modern science, soil is also alive, or rather, made of life. Many of the same trillions of bacteria, fungi, and microorganisms that live inside soil also live inside us.

Each human being has at least as many microorganism cells as human cells.[1] Scientists tell us that those microorganisms were the beginning of life on Earth. They must have liked it here because they never left. These little critters are the keys to our health and the keys to our food. They connect us to the soil.

Without microbes we have no soil life. Without soil life we have no food. It should go without saying that without food we cease to exist. This primal relationship is our most fundamental contract.

Perhaps that's why for centuries, millennia even, humans have held the soil in high regard. From the Sumerian deity Emesh whose domain was vegetation, to the Egyptian god Osiris's realm of the underworld, rebirth, and grain, to the Greek Gaea (Earth) to Kokopelli in Native America to Jesus's "Parable of the Soils" in the book of Luke, ancient humans have revered and cherished their link to the soil that sustained them.[2]

Ours is possibly the first large civilization to have no particular mythical representation (and reminder) of the importance of soil. Granted, one could argue we don't have mythical representations of the sky or rain or other natural phenomenon. But we have no problem worshiping our own forms of entertainment, technological, and corporate deities. Ours is a neo-mythical view of the world in which a light switch magically produces light, a car magically drives, and food magically appears on grocery store shelves.

The farther away from the source, the easier it becomes to forget where all our stuff comes from. It doesn't matter, though. Because ultimately we are nothing without the soil. This is a big realization. It's so big that it's a hard connection to make when eating a bag of Doritos, drinking a Coke, or eating a Big Mac. Despite the fatty, salty, sugary, colorful, perfectly packaged, and easily bought world of our food today, soil is making a comeback in pop culture.

Three of the recent movies up for Oscars had soil as a pivotal plot point. In *Interstellar*, humanity has destroyed the world's soils, causing a global dust bowl. It's up to Matthew McConaughey to save us by finding another planet, one with its soil still intact. In *Mad Max: Fury Road*, the soil has also been killed, and it is Charlize Theron and her band of siren-like women who ultimately overthrow an oppressive male-dominated dictatorship to bring water and seeds back to life. If those two weren't enough, Matt Damon makes agronomy cool when he uses human feces as fertilizer to bring the dirt of Mars to life in order to become the first off-planet farmer in *The Martian*. It seems as if the storytellers of our time are trying to tell us something.

These stories are part of our modern mythos. It turns out that for all our gadgetry we are still bound by the same contract as our hunter-gatherer ancestors. Soil gives food and in so doing soil gives life. We can go as far away from the source as Mars. But ultimately, Earth is our home and without the substance we call earth, we cannot survive.

This is why we humans will tell this same story in some form or another over and over again. It's why we show our children how to garden and how to farm. And it's why they show us how to eat dirt. Us and soil—it is a time-tested relationship. No matter how often we leave each other, we will always come back together.

The Sufi poet Rumi, who was no stranger to drought, is quoted as having said: "There are hundreds of ways to kneel and kiss the ground; there are hundreds of ways to go home again."

The ground, otherwise known as the soil, is our home. No matter where we go, or how far we venture, or how long we live, we will always return home.

And because of this, and because of the discoveries in the pages that follow, I have great hope.

CHAPTER 1

SHOWDOWN IN PARIS

There is something magical about wintertime in Paris. The cold makes the stone older. The rain makes the coffee better. Hugs are warmer. Life is more precious, especially this winter.

Humans have lived in this place for almost ten thousand years, since soon after the ice sheets of the Pleistocene Epoch retreated. In many ways, the survival of this city is a testament to French agriculture, which has been practiced on some of the same earth in lands surrounding this city for millennia. It is strange to think that it may have been the Phoenicians who established the foundation for French cuisine.

Unlike so many ruined relics of once great cities past, this metropolis is a thriving tribute to the human spirit. Its ancient grandeur is somehow only enhanced by its chaotic modernity. It is called "La Ville-Lumière" (The City of Light) because it was an early center of ideas and education in Europe (and also one of the first cities to adopt street lighting). But wherever there is permanence, the ephemeral vanishes as quickly as it appears. And wherever there is light, darkness lurks.

Only days before my arrival on Parisian soil, a series of horrific shootings took the lives of 130 civilians and injured hundreds of others. The group claiming responsibility for the attacks is called Daesh, otherwise known as ISIS/ISIL, the terrorist organization that originally emerged in Iraq and now has a new base of operations in Syria. The men responsible for the attacks were primarily Belgian and

French nationals of Syrian descent. Many of them had been lured into the organization by jihadist recruiters based in Europe.

But one could say all of this really actually began with water or rather, the lack thereof.

The most recent drought in Syria began in 2006 and has gotten progressively worse as average annual rainfall has dropped to below eight inches, the "absolute minimum" to sustain unirrigated farming. Farmers tapped thousands of new wells but the water table in the region's aquifer quickly dropped. Crop failures in some areas reached 75 percent and as much as 85 percent of their livestock died of thirst or hunger. Farmers fled to the cities en masse.

Syria's internal refugee count, as well as its poverty, skyrocketed. Hunger soon led to riots, which led to a bloody civil war, which forced four million people from their homes. Meanwhile the floodgates were opened for any and all forms of gunmen, freedom fighters, and terrorists to enter Syria and attempt to wrest control of the fractured state.[1] Anger, hunger, desperation, weapons, and millions of homeless people—Syria was a powder keg waiting for a match. It was also a fight looking for an enemy, which ISIS readily found in the West.

This makes me thankful for the chilly, misty precipitation that now coats my face as I step out of the bus at Le Bourget, the former primary airport for Paris. I am also thankful for the countless beret-donning, heavily armed soldiers who stand imposingly everywhere I look. I am not at this old airport to catch a plane, however. Instead I am here to observe an announcement that will happen at this year's United Nations Framework Convention on Climate Change, also known as COP (Conference of the Parties).

Paris is covered in signs for COP21, and the friendly green vest-wearing young people who notice my badge smile and say "Have a nice COP!" which is both amusing and endearing. The parties of the COP are the 195 participating nations of the world. The "21" designates that this is the twenty-first time they have met in an attempt to find solutions to the looming threat of higher quantities of greenhouse gases in the atmosphere.

Along with some of the thirty-eight thousand diplomats, businesspeople, press, and members of "civil society" who are official attendees of this year's event, I make my way through the TSA-like security and into COP. This being a former airport, the events are held inside massive buildings that were once hangars. A large cement breezeway serves as the home for media stands as well as shops with coffee and croissants (it's France, after all). Because I am here to interview a member of the French government, I have managed to score a diplomatic badge, which gives me access to areas that are off limits to the general public and press. Using my credentials, I stroll into one of the "Blue" diplomat zones.

The interior is set up like a showroom floor for a Las Vegas–style convention on international tourism complete with various countries advertising their best travel locations. Large boxlike booths are arranged in a grid, forming a kind of pavilion city with streets running lengthwise and crosswise. With a few notable exceptions, the displays seem like they were cobbled together at the last minute by grade-school kids.

At the Korean pavilion there's the "take your photo by sticking your head in the cutout of a traditional Korean-dressed woman or man." At the US pavilion (aka US CENTER 2015 PARIS), which is set-dressed like a political convention stage complete with red, white, and blue banners, there is a "take your selfie here" spot that looks like a US identification card. As I walk around, I see a smattering of half-hearted displays on solar, deforestation, and the like. The effect of the show is somewhere between underwhelming and farcical.

Outside the Chinese pavilion I am halted by a robot that barks at me (in loud robot English) and asks me where I flew from. I tell it that I came from "Loooos Aaaannngeles," making sure to carefully pronounce the city's name. Unfazed by my answer, it says, "Touch my screen" (located on its chest). When I finally manage to key in "L-A-X" onto the Teletubby-like bot, it loudly announces how much carbon I put into the atmosphere by flying to Paris (rather than how much CO_2 the coal-burning, iPhone-producing nation of China emits annually).

On my way to find a croissant I stumble upon the tired "Let's get four people together to ride these stationary bikes to light up this display" display. Three younger female Asian diplomats in matching outfits madly pedal away in an attempt to sustain the lights while an older gentleman appears to be asleep at the fourth bike. I am now completely turned around and succumb to wandering the hallways in hopes of finding an exit. I soon find the big-money pavilions. These lavish displays of nonsensical eye candy have nothing to do with reducing carbon emissions and everything to do with purposeful confusion. The larger the emissions profile of a country, the more money seems to have been spent on their display.

India, the third-largest emitter of greenhouse gases behind China and the United States, has obviously laid down some serious euros for their display. It's a marriage of a Bollywood movie set and what appears to be a Brookstone store. Six Indian men dressed in what looks like traditional Indian wedding attire stand like stuffed sentries around its exterior. Fiber-optic lights pulse, screens display all sorts of images, diplomats around the edges stare in awe and take photographs of the spectacle. The only thing that would make it more interesting would be a spontaneous line dance by the sentries, but beggars can't be choosers.

Then I come across my favorite COP21 display, "The Gulf Cooperation Council Pavilion." It's not referring to cooperation by the states around the Gulf of Mexico, but rather the other Gulf, the big one in the Middle East. It is, by a factor of 2x, the largest pavilion. On the outside its vaguely Arabian latticework brings the eye to a map of Saudi Arabia framed inside a bold hexagon logo. Inside, marble-esque walls meet polished hardwood floors to give that very empty, high-end retail space vibe. Its only prominent interior feature is what appears to be a longer than average state-of-the-art air hockey table. Upon further inspection I find this thing is a touchscreen table designed to advertise the efficiency of oil operations across the Gulf states. None of this changes the fact that if I stumbled into this pavilion after the show's requisite cocktail hour, I might think somebody had stolen all the computers from the Dubai Apple store.

They say that how you do something is how you do everything.

At least when it comes to the United Nations dealing with the issues of greenhouse gases and climate change, that is certainly true. To understand how we got to the point of Teletubby bots, selfie spots, and touchscreen air hockey tables at what is supposed to be a conference to decide the fate of the world, we need to go back to the 1950s.

THE AIR UP THERE

Charles David Keeling was an avid outdoorsman and concert pianist. He was also a scientist with a keen interest in the environment. Keeling wanted to know if humans were affecting the level of carbon dioxide in Earth's atmosphere. So in 1955 he built an infrared gas analyzer capable of measuring CO_2 concentration.

Working with Scripps Institution of Oceanography and with the United States Weather Bureau, Keeling established measuring stations in Antarctica (close to the South Pole), on the Big Island of Hawaii (at the Mauna Loa Observatory), and in La Jolla, California, all of which were away from the interference of city pollution. In 1960 Keeling published his first findings in an unassuming paper. This report sent shock waves through the scientific community.

Keeling made two important discoveries. The first was that overall, carbon dioxide emissions in the atmosphere were increasing. In just three years, between 1957, when his measurements began at the South Pole, and 1960, and when he published his report, CO_2 increased from around 310 parts per million to around 315 parts per million. While some scientists had postulated for a long time that humans could, through their activities, increase CO_2 concentration in the atmosphere, this was the first absolute proof.[2]

Keeling's second discovery was completely unexpected. When his sampling data was plotted onto a graph, it began to resemble a curve, not just an upward curve but an oscillating one. The curve went down in the summer and came back up (higher than before) in the winter. This "Keeling Curve," as it became known, indicated something amazing—in very general terms, the planet is breathing.

More accurately stated, it is the immense quantity of living plant material on Earth that breathes (otherwise known as respiration). During the spring and summer, as plants and crops grow via photosynthesis they use sunlight to convert carbon dioxide (CO_2) into plant tissue, leaves, seeds, roots, and root exudates (carbohydrates and sugars). Then in the fall and winter, as trees lose their leaves and plants go to seed, die, and decompose, their carbon is oxidized (combined with oxygen to form CO_2) and goes back into the atmosphere. This was well understood, but nobody had ever seen the effects of this biological process on a global geochemical scale.

Thanks to Keeling, there it was in black and white—respiration. Plants, it turns out, and the microbial community that supports them, have the ability to pull so much CO_2 out of the atmosphere at one time as to make a measurable dent in the global concentration level of CO_2. For the past six decades, this discovery seems to have been

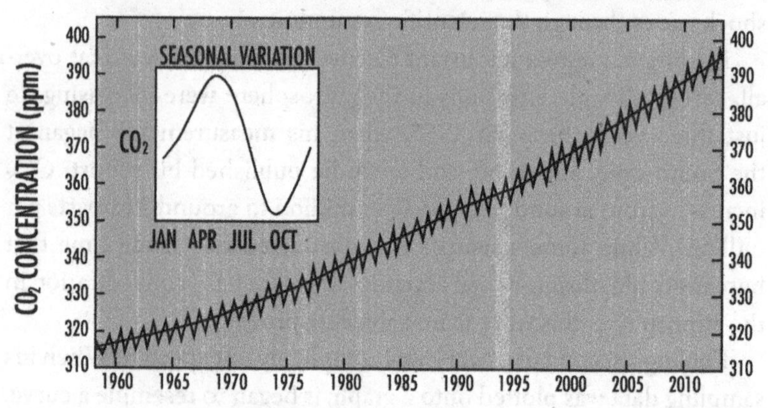

Atmospheric carbon dioxide measurements from air samples taken at the Mauna Loa Observatory. The close-up (white box) shows the "respiration" of Planet Earth. The rise in the curve is from plants dying in the winter and releasing CO_2. The drop is from plants growing in the spring and sucking up CO_2.
Data source: NOAA.

an open-and-shut case to the scientific community, kind of a ho-hum type of thing.

But this is the very discovery that could prove pivotal to stabilizing atmospheric CO_2 and growing a lot more, and a lot better, food.

Since Keeling's first measurements, CO_2 levels in the atmosphere have continued to rise. Today the atmosphere contains about 400 parts per million of CO_2. Apparently none of the prior United Nations COP meetings have slowed the rate of CO_2 increase. If anything, accounting for the hundreds of thousands of flights and car trips needed, the COPs probably increased CO_2 levels. It also seems that no lasting change took place as a result of previous COPs, most of which ended with agreements to agree or decisions to decide, or ways to do something later.

This brings us back to Paris, and to the twenty-first COP. It is here that the parties will "lay down the gauntlet" and make a "firm and lasting decision" on how to deal with this big man-made CO_2 monster once and for all. Yes, they will also do some dining and no doubt some drinking and there will be some business done as well. Among those present will be citizens, trillion-dollar industries, billionaires, and millionaires who stand to gain if CO_2 stays the same (or goes up), and those who stand to gain if it goes down. On all fronts, pressure will be applied, humans being human and corruption being what it is.

There will be nonprofits working hard to leverage their research and their membership base to try to make headway on the issues of poverty, deforestation, renewable energy, and the like. As a result of the general feeling that the public is being shut out, there will also be a blizzard of protests, media attention–grabbing actions, and a nearly endless array of side events.

But I'm not here for any of that. Instead I'm here to meet the one man who might actually have a solution to rising atmospheric CO_2 emissions. And his solution has nothing to do with the global warming debate and everything to do with food.

À L'HÔTEL

I am standing in the early-morning dew on the street outside the home of the prime minister of France. It's my second time talking to the guards, who previously told me to go wait down the street and get a coffee (in other words, go away). As I am trying to explain, I do have an appointment. But my American passport and broken French aren't helping.

Finally, my liaison arrives and, to my relief, is able to explain the nature of the meeting to a more official-looking gentleman who gets yet more security personnel, and at long last I am ushered inside with a horde of well-armed men. Another passport check, a few more phone calls, another security screening, and I am let into the courtyard of the Hôtel Matignon. Constructed between 1722 and 1725, its stone edifices form a stunning 360-degree enclosure fit for a king.

I stroll to the far side of the courtyard and admire the inscription at the top of one of the stone walls. It is the Declaration of the Rights of Man and of the Citizen, the document that Thomas Jefferson helped General Lafayette draft at the start of the French Revolution.

Without warning the large portcullis to the courtyard shoots upward, the anti-vehicle barrier of metal posts drops, and a swarm of flashing lights and vehicles screeches into the parking lot. Police and security spread out everywhere. Between two large dudes in flak jackets who have unsubtly positioned themselves directly in front of me, I can see a smallish, balding man get out of a vehicle. It's François Hollande, the president of France. It's all very James Bond (or OSS 117, as the case may be in France), but he's not the guy I'm here to see.[3] He darts inside the building, and the security detail disappears as quickly as it arrived.

I am told "my man" will arrive next. I brace myself for more flak jacket–wearing Frenchmen. I wait. I am offered a coffee. I decline. I wait some more. It's getting wetter and colder. Finally, just as I'm about to ask for that coffee, the portcullis shoots up, and, unceremoniously, in strolls a tall, confident man in his midfifties. He's also got a security detail, two thin, sweaty dudes in suits with earpieces. They look winded from the walk. They're not very intimidating. But then again, nobody has ever tried to assassinate the French minister of agriculture.

Standing about six-foot-two, with a full head of well-styled gray hair and a permanent smirk, Stéphane Le Foll has an air of infectious confidence and a Bill Clintonesque charisma. He shakes my hand, welcomes me to France, gives me a once-over, and motions for me to follow him as he struts full speed toward the building. This is a guy who doesn't waste time.

As we walk he asks if I need a coffee. I thank him and assure him I'm good. I follow Le Foll into a beautiful office overlooking a park. There's a long table around which several well-dressed men and women sit poring over a pile of documents. Le Foll takes a seat and quickly pages through a newspaper that has been set out for him. He asks his team for an update. The people at the table take turns speaking about carbon numbers for various countries.

He puts the newspaper down and asks a few pointed questions. He is double-checking their work, not in a condescending way but in a way that shows he knows what they are talking about. This group of men and women is assembled from the French National Institute for Agricultural Research (INRA), the top agricultural research institute in Europe and one of the world's premier centers of agricultural science. Established in 1921 and formalized by law in 1946, INRA

Stéphane Le Foll is a man on a mission. (*Simon Balderas*)

has more data on soils and food than any organization save the US Department of Agriculture.

This team of advisors is not discussing the carbon emissions from each country. Instead they are calculating how much carbon each country can store in its soils. The huge pile of papers on the antique table represents the first global analysis done by a government on the carbon sequestration potential of humankind's agricultural soils.

And according to what the team of IRNA scientists gathered around the table has surmised, the potential is staggering.

THE "SOILUTION"

According to Dr. Rattan Lal of Ohio State University, a pioneer in the study of "biosequestration" (using plants and microbes to sequester carbon dioxide), humans have put some 500 gigatons (billion tons) of carbon dioxide into the atmosphere since the birth of agriculture some ten thousand years ago. But most of that CO_2 was emitted during the relatively recent advent of modern agriculture. Through plowing the land, which releases tremendous quantities of CO_2, deforestation, urbanization, and land-use change we have effectively taken a massive quantity of carbon that used to be stored in the ground and released it into the atmosphere.

In addition, about 250 years ago humans began to burn copious amounts of fossil fuels. That added another 350 gigatons of CO_2 to the atmosphere (half of which have been produced just since 1980).[4] Total it all up and we've put some 850 gigatons of CO_2 into the air. For now, the majority of CO_2 humankind has released since the birth of civilization has come from plants and the soil.

Agriculture Minister Stéphane Le Foll has a simple plan: put that carbon back where it came from. And instead of taking ten thousand years to do this, he wants to do it in as little as two to three decades. It's a radical, bold, elegant, and disruptive idea. And according to the math his INRA advisors have been doing, it just might work.

The basic science of biosequestration has been understood since

Charles Keeling did his first measurements of CO_2. But it was assumed that land and plants would always emit the CO_2 they inspired. In other words, the net CO_2 uptake of land would always zero out. But a recent and growing chorus of biologists is saying something quite different. Their assertion, based on test plots in places like France, Ohio, Pennsylvania, and Australia, is that when properly managed, agricultural land can be a net "sequesterer" of CO_2. Meaning more CO_2 in than out. This is a very big assertion.

The biology involved in biosequestration is complex and has heretofore been all but dismissed by the climate science world (which has thus far been limited to the atmospheric, oceanographic, and chemical fields). The fields of science are like silos. Chemists and biologists generally do not mix. Physics and chemistry being predictable and biology being infinitely less predictable, the science of biosequestration has been slow to be understood and slower still to be accepted by the climate change science community.

In general terms, carbon is present in all living things. To grow plants, or crops, or food, the more bioavailable carbon there is in the form of the dark, rich topsoil, otherwise known as humus or soil organic matter, the better plants will grow. For a long time, scientists have known that as plants and animals die and decompose they can add to the topsoil. But as higher-resolution microscopes and more accurate instrumentation became available, biologists have been able to look beyond the macro layer of life (plants, leaves, roots, etc.) and into the micro layer (fungi, bacteria, etc.). That's where the real carbon action is.

Plants, through their roots, excrete carbon in the form of root exudates. These exudates feed trillions of microorganisms, which are involved in extremely complex biological and chemical exchanges with the root systems. In a working ecosystem, carbon excreted from roots is carried through a series of handoffs through the upper, pliable "labile" layer of soil as it moves into the deeper, more immobile layers of the soil. It is eventually deposited in the form of organo-mineral complexes deep within the recalcitrant fraction of the soil. That's where it can stay for thousands of years.[5]

THREE FUTURES

How much carbon can be stored? Well, that depends on who you talk to. Agronomists in Le Foll's camp, as well as the ones studying this issue in the United States and Australia, believe that widespread implementation of "light" regenerative agriculture techniques like reducing soil tillage and planting cover crops could cause the same amount of CO_2 to be biosequestered as we humans currently emit each year. I call this our "medium-case" scenario. ("Worst case" being we do nothing and hope the UN will solve this problem for us with more COP meetings.) This soil-based, medium-case biosequestration scenario still necessitates that we ratchet down the world's use of fossil fuels, and it is markedly better than our current scenario of continuously adding more CO_2 to the atmosphere. But not for the reasons being put forth by the current climate change community—i.e., "saving the polar bear" or "stabilizing global temperatures." While these crusades are important, there is a far more pressing issue.

Simply put, when we add carbon dioxide to the atmosphere, Earth's natural balancing mechanisms attempt to remove it. The place where it is most readily absorbed is in the oceans. In the oceans, CO_2 turns into carbonic acid, causing the pH of the oceans to become acidic. As the oceans acidify, the basis of the food chain—phytoplankton—dies.

This may sound bad, because we tend to like to eat things like fish and lobster that rely on many smaller layers of food creatures. But that's not why phytoplankton are important to this discussion. These tiny, delicate creatures produce more than 50 percent of the oxygen we breathe. Thus more CO_2 in the atmosphere equals acidic oceans, which equals dead phytoplankton, which equals insufficient oxygen for humanity.

People on both sides of the global warming debate have completely missed this pivotal truth: It is irrelevant whether or not increased levels of CO_2 in the atmosphere cause warming if we cannot breathe. Stately differently, if we humans wish to live in a flourishing world well

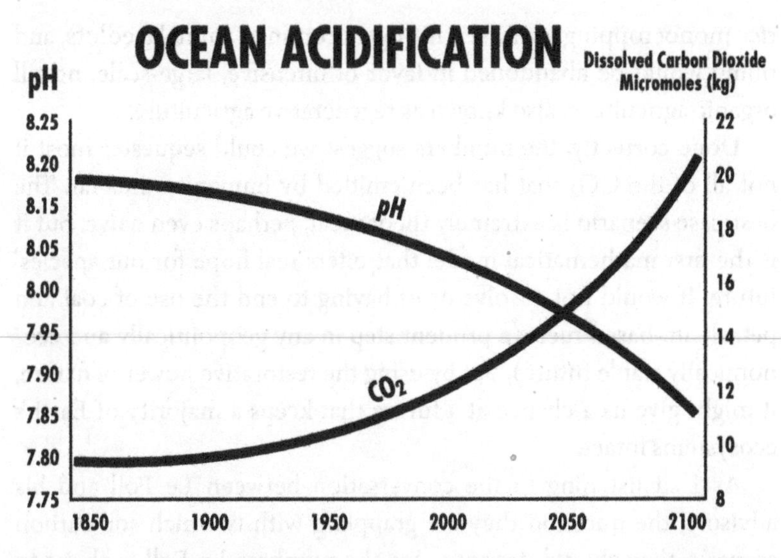

As carbon dioxide in the atmosphere rises, the oceans absorb as much as possible, turning the CO2 into carbonic acid. Ocean acidification threatens phytoplankton, which produce half the oxygen we breathe.
Data source: NOAA.

into the future, then pulling the CO_2 back down into the ground is as important as the air we breathe.

The medium case of drawing down the CO_2 we emit yearly would theoretically buy humanity time to wean itself off its insatiable thirst for carbon-based fuels. But even if we stabilize CO_2 at 400 ppm, scientists tell us the Earth as we know will undergo dramatic changes. The oceans will continue to acidify, exponentially more species will be lost, and human civilization will be threatened. It's better than the "worst-case/hope the governments of the world will work together to figure this out" scenario, but it's still bleak.

However, there is a "best-case" scenario. Granted, it requires rethinking global agriculture at its core. It may even involve rethinking the idea of food itself. But the best-case scenario is palatable, if not actually good. In this scenario, an aggressive global program would have to be instituted to reform agriculture so that big moneymakers like herbicides, pesticides, genetic engineering, corn, soy, wheat, and rice monocropping, synthetic nitrogen, confined animal feedlots, and tilling would be abandoned in favor of intensive, large-scale, no-till organic agriculture, also known as regenerative agriculture.

Done correctly, the numbers suggest we could sequester most if not all of the CO_2 that has been emitted by humanity thus far. The best-case scenario is extremely theoretical, perhaps even naïve, but it is the first mathematical model that offers real hope for our species' future. It would not absolve us of having to end the use of coal and petroleum-based fuels (a prudent step in any geopolitically and economically stable future), but by using the restorative power of nature, it might give us a chance at a future that keeps a majority of Earth's ecosystems intact.

As I sit listening to the conversation between Le Foll and his advisors, the question they are grappling with is which soil-carbon scenario they should propose. Are the numbers Le Foll is about to announce at COP conservative enough to be achievable? Based on their extensive research, the team believes that they are. Should he be more aggressive and propose a best-case scenario? The question is

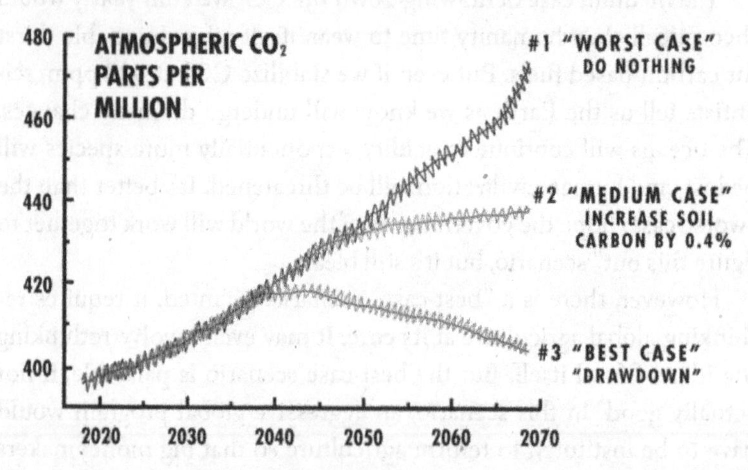

An approximate rendering of our three possible CO_2 futures.

met with a moment of silence that the minister breaks by saying, "We must first get an international agreement. There will be time to rethink our position once a coalition is under way and we see results." He lets that sink in for a moment. Nobody says anything. "Very well," says the minister, "shall we go?"

And with that he is off again, walking at breakneck speed toward his van.

DANS LES RUES DE PARIS

I squeeze into the small minivan with Le Foll, his security detail, and some of his entourage. Still more security men jump into a lead car. Le Foll has managed to keep his newspaper, which he attempts to have another look at as the vehicles leap past the raised portcullis and hit the cobblestone streets with aggressive speed, sirens blaring. As Paris zips by outside, Minister Le Foll puts down his newspaper with a note of frustration and instead of reading the daily headlines, slowly unveils his plan for attempting to save the world. "*Bon*," he says, "where shall we begin?"

I ask him how he came up with this idea of soil-carbon storage. Le Foll says, "The question you're asking necessitates you understand that the first tool in agriculture is the soil, not the tractor, not the combine harvester, not satellites, or computers, it's the soil. And this is very, very important because we can also reduce greenhouse gases by carbon storage in soil." The minister explains that the fact we have forgotten how agriculture works in the first place is the reason why soil-carbon storage seems like an announcement at all. He's not, he asserts, saying anything new, but rather putting very old knowledge into a new global context.

Satisfied he has solidified the importance of soil, he tells me, "I've been working at this for ten years, since I was in the European Parliament. I knew that in the US there were researchers and farmers working on the question of carbon storage. There was even a proposal made by Bill Clinton, a document that examined the potential for carbon storage in the soil in the US, but it was never realized." Le Foll distinguishes that the research is important, but only if it alters global policy, which is what France is trying to do.

The minister says that for years France has been working on a new model of agriculture that brings together the best of old-world knowledge and the best modern practices. It's called "agroecology." He explains the model as follows: "Agroecology is the combination of three elements: economic performance, ecological performance, and social performance. And behind these, there are, and will always be, farmers. That's what has been built into the French concept. I am working on a law that says 'The environment is not the enemy of agriculture, and agriculture is not the enemy of the environment.' It's just the opposite. The two are totally connected."

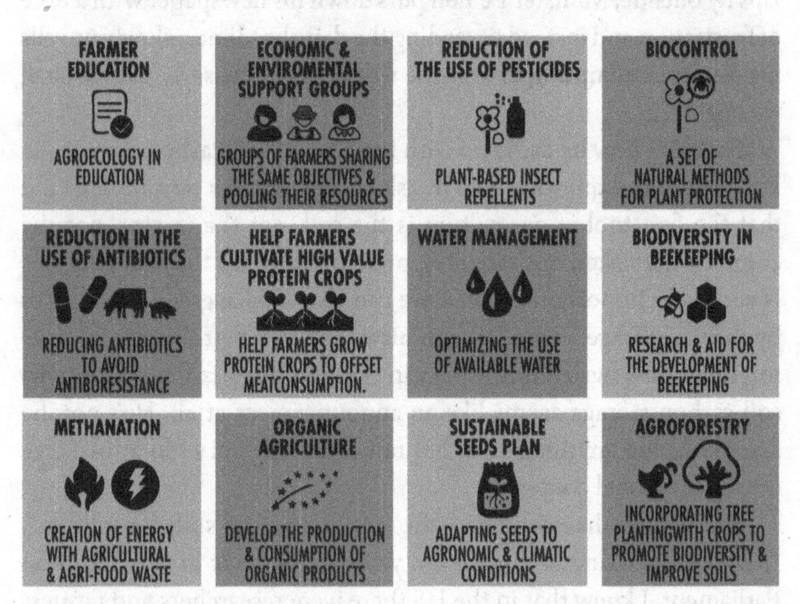

France's Agro-Ecology program is the most developed governmental regenerative agriculture program in the world.
Source: INRA.

This sounds nice but it seems a bit "Let's hold hands and sing Kumbaya." After all, the goal of agriculture, as we are told ad nauseam in the United States, is to feed the world. So I ask Le Foll how he intends to do just that—feed the world with agroecology. He takes a breath and launches in. "In the month of October we were in a field after a wheat crop had been harvested. That field had beans, oats, and alfalfa more than a meter (three feet) high. In the month of October! Over a meter high! And there we are standing in it and it's almost winter. Ten years ago no one would have imagined this was possible."

The minister says that their test fields show that through agroecology, farmers can produce significantly more food per acre. He says that the secret to growing more is to increase the carbon content of the soil. And to do that, he says, one must keep the soil covered with vegetation and never leave it bare. He's fired up about this concept and his answers are becoming more emphatic. He explains that "we're going to see very technical farmers who store carbon in the soil, develop more food production, keep crops on planned rotations, and save biodiversity all at the same time. This has huge implications for global agriculture." If what he is saying is true, it also has huge implications for feeding the world.

Our van has been stuck in traffic for some time when Le Foll finally looks at his watch and says to the driver, "Gibbon, we're going to be late." (In other words, do something.) Just then, we hear sirens and engines revving from behind us and a diplomatic convoy screams by going the wrong way down the opposite side of the street. In an instant, Le Foll's driver begins talking into his radio as he turns the wheel and punches the gas. The van lurches forward, becoming part of a convoy of loud sirens and chasing vehicles. As the engine whines loudly, I can't help but glance at the speedometer a few moments later. It's well above 100 kph (60 mph) and accelerating. We merge onto a thoroughfare and are now part of some sort of quasi-diplomatic group of vehicles *going the wrong way* down a freeway. The security detail all speak on their earpieces. The tension inside the van is palpable.

Le Foll appears only momentarily disrupted. For him, it's just another day at the office. He looks at me and says, "Well, what else?" I need him to explain how he came up with the program he is announcing today at COP21. In his press release (which was confusing even to my French-American friends who helped me translate it) it is called the "4 for 1,000" program.

Speaking loudly over the sirens, Le Foll patiently explains the basic math. Each year, humanity emits 4.3 billion tons (gigatons) of carbon into the atmosphere. Meanwhile, the world's soils contain 1,500 gigatons of carbon as a fixed stock. Thus, if we can increase the carbon in

4 FOR 1,000

CARBON SEQUESTRATION IN **SOILS** FOR **FOOD SECURITY** AND THE **CLIMATE**

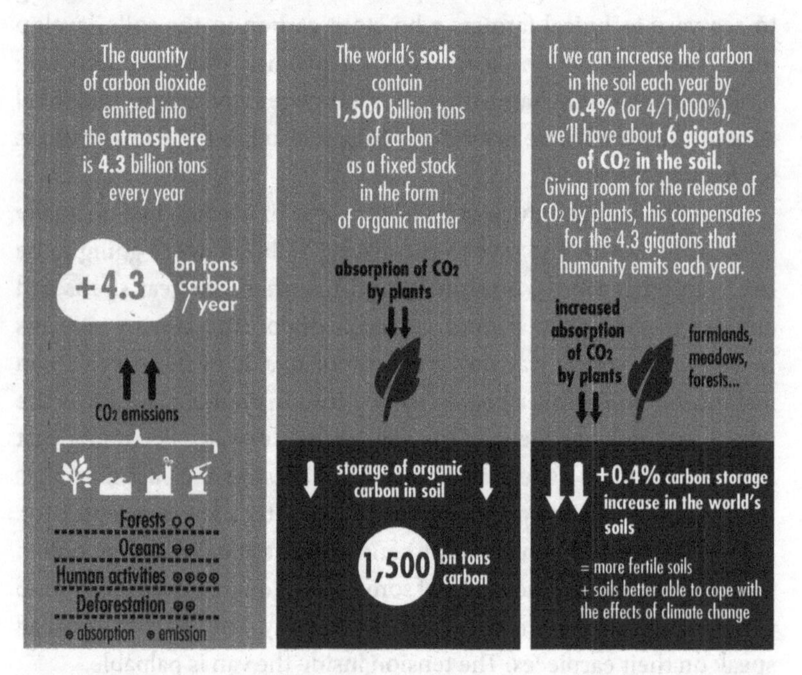
The quantity of carbon dioxide emitted into the atmosphere is 4.3 billion tons every year
+4.3 bn tons carbon / year
CO_2 emissions
Forests
Oceans
Human activities
Deforestation
absorption
emission
The world's soils contain 1,500 billion tons of carbon as a fixed stock in the form of organic matter
absorption of CO_2 by plants
storage of organic carbon in soil
1,500 bn tons carbon
If we can increase the carbon in the soil each year by 0.4% (or 4/1,000%), we'll have about 6 gigatons of CO_2 in the soil. Giving room for the release of CO_2 by plants, this compensates for the 4.3 gigatons that humanity emits each year.
increased absorption of CO_2 by plants
farmlands, meadows, forests...
+0.4% carbon storage increase in the world's soils
= more fertile soils
+ soils better able to cope with the effects of climate change

The aim of the 4 for 1,000 Initiative is to demonstrate that healthy agricultural soils can sequester CO_2 while providing more food per acre.
Source: French Ministry of Agriculture.

the soil each year by 4/1,000 percent (or 0.4 percent), we'll store about 6 gigatons of CO_2 back into the soil. Leaving room for the release of CO_2 by plants, this compensates for the 4.3 gigatons that humanity emits each year.[6]

If the math sounds funny, it is because we generally think of a percentage as shorthand for a numerator over a denominator. In America we might say "4 percent" whereas the French might say "4 for 100," which just means 4/100, which is the same thing. So "0.4 percent" can be expressed as "4/1,000" and can also be stated as "0.004." To make the translation into English even more confusing, the French use the special numeric symbol ‰, which means "parts per thousand." To the American eye 4‰ might easily be mistaken as "4 percent" but that's not what it means at all. It means "four parts per thousand." (You say "tomaaaato," I say "tomahhhto." Same difference.)

The bottom line here is that the French minister's program, which he calls 4 for 1,000 (written as "4‰"), is just a European way of saying that in order to offset annual global CO_2 emissions, we need to annually increase the soil carbon of the 5 billion hectares of agricultural land on Earth by an average of 0.4 percent, a small but significant shift toward sequestering CO_2 back where it came from. It's a lot to wrap your head around, especially when flying down tight roads across Paris in a van filled with heavily armed men.

Le Foll says the 4 for 1,000 project has significant co-benefits. "At the same time that we're fighting against climate change, we are also assuring the fertility of the Earth's soil and therefore assuring agricultural production to feed the planet," he says. I ask him if this doesn't go against the grain (so to speak) of the major global push toward more industrial types of agriculture. A knowing smile flashes across his face.

Le Foll checks his watch again and looks out the window. Satisfied with our breakneck speed he explains, "In the first Green Revolution we used machinery and chemicals. But we went too far. If I may use your country as an example, the use of genetically modified organisms (GMOs) was presented in the US as a solution to solve the massive problem of world hunger. The claim was these GMOs would require less chemical products. The result fifteen, twenty years later is that

GMOs increase the use of these chemicals. Therefore, this is not what is going to solve the problem of hunger in the world. So, I am convinced that in the Monsanto agricultural model of GMOs and their widespread use the herbicide glyphosate cannot feed the world. We have arrived at the end of that model. We are at its very end." Glyphosate, by the way, is the primary chemical component of Roundup, a poisonous herbicide discussed more in chapter 2.

Le Foll tells me that Monsanto and their allies have not given up. They are presenting their own model of agriculture at COP, which directly competes with the 4 for 1,000 initiative. They call it "climate-smart agriculture" and, not surprisingly, it's being promoted by a Monsanto-backed "nonprofit" consortium called Global Alliance for Climate-Smart Agriculture (GACSA for short). Essentially, GACSA is promoting patented GMOs as a way to solve the climate crisis. Even though the chemical industry's own studies show no added soil carbon as a result of using GMOs and their required chemicals, they assert their methods can in fact be "regenerative" to soil. Says Le Foll, "If we were just to stick with climate-smart agriculture, I don't believe we can even have a real conversation about the future of agriculture."

Not surprisingly, this is the model that the head of the USDA, Tom Vilsack, is supporting.[7] Le Foll tells me he is meeting with Vilsack the next day to try to convince him to come over to the 4 for 1,000 side. (In other words, "Leave the Death Star and join Han Solo and the Rebel Alliance.")

Our van has left the freeway and merged into a much larger group of police motorcycles and escort vehicles which, because most of the streets are blocked by police, is the only traffic going anywhere. Luckily for morning commuters, the Métro system is moving like clockwork.

As we pull in through the security checkpoint, I ask Le Foll if he is nervous. "Nervous? No. Why would I be? I feel, how do you say it? Motivated."

AN AUSPICIOUS ANNOUNCEMENT

We pile out of the van. The cool, damp air is refreshing, especially after so much sweat and adrenaline. Luckily for the minister and several of his staff, we have made it, just in time . . . to have a cigarette.

For all their progress in so many other areas, the French are still smoking plenty of cigarettes. But even that has gone down in recent years with smoking bans in numerous public places, including (thank goodness) the Métro.

Just when it feels like this break may last long enough to begin checking smartphones, an attaché comes up and firmly announces, "It's time." Cigarettes are quickly extinguished and once again Le Foll is on the move. Like the rest of his gang, I find myself walk-jogging after the minister. We are waved through security at a back door and the entourage somehow expands to some twenty advisors, more security personnel, and various other VIPs as we all walk-jog through the various COP international tourism pavilions.

Unlike the lavish Indian and Gulf states' pavilions, the French one is an unimpressive, unadorned wooden box with a few fancy blue stage lights, just barely big enough to seat two hundred people. By the time we walk in with the minister some three hundred bodies are crammed into the little presentation room and it's standing room only. Le Foll is ushered into a seat in the front row. I am motioned to stand awkwardly against the sidewall.

Judging by the sweltering heat inside the room, the program has been going on for some time. National representative after national representative is brought up to the microphone to voice his or her (but mostly his) country's commitment to the 4 for 1,000 program. There is Germany, followed by a smattering of African countries, then Belgium, Australia, New Zealand, and many others. They're all on board. After each person speaks, they walk over to a small camera-ready podium and sign an official 4 for 1,000 commitment document, wherein they are committing to transition the entirety of their agriculture to this program. As they put down the pen, each one is photographed. There's no backing out for these folks now.[8]

While a diplomat from some former Soviet republic gives a long Tolstoyesque speech, I notice several glaring omissions from the map of signatories that is projected onto the wall behind the speech-givers. India, China, the Gulf states, and America are all grayed out. In other words, none of the planet's biggest CO_2 polluters will be signing the 4 for 1,000 document today. I check the crumpled program in my pocket; not one person from the United States is even speaking. The US pavilion with its red, white, and blue set dressing is only steps away, but somehow the greatest agricultural producer in the world couldn't bother to even send an emissary to thank the French for spearheading this program.

Finally, it's time for my man to speak. The room perks up as Le Foll strides to the podium. He grips the clear plastic of the podium so tightly that I think he might crush it with his bare hands. But those same hands are soon animated in the air.

Le Foll begins quietly and says:

> Ladies and gentlemen, my dear colleagues, it was here, just a few weeks ago, that the city was besieged by terrorist attacks. To hold this Climate Conference in Paris is to hold the idea that humanity must be capable of mobilizing itself, of uniting and of bringing hope. We must be able to realize that together, all the men and women of this planet, of every country everywhere, that we have the capacity to build the future, and to do so in brotherhood and in peace.
>
> This 4 for 1,000 initiative makes me proud. Not proud to have started this, not because this initiative was conceived here in France. Proud because it has brought forth awareness. The idea of this ratio, 4 for 1,000, is that with the agricultural soil, and also for the soil of the forests, we have enormous potential. A capacity to solve, and this is the hope, the giant challenge in front of us with the fight against climate change. That's colossal. And that is the potential we must strive for in order to survive.

> But to solve the question of the storage of carbon, it's necessary that we have accompanying public policies. And we cannot succeed without farmers. Our farmers started this a long time ago. It was our Brittany grandfathers who said, 'There are three elements to agriculture: manure, manure, and manure.' So this idea of working with nature has existed since the beginning of agriculture here. We are going to better utilize natural mechanisms before going against them in the development of our agricultural production. In France we call that agroecology, and it's going to utilize a lot more of what nature is able to offer us so that we may produce more, and produce it better.
>
> This is not only about self-preservation. Rather, this is about the consciousness of the future of humanity. Four for 1,000, that is the potential. Four for 1,000, that's the goal of this grand project. Four for 1,000, that's the ability we all have together and how we will be able to succeed. Thank you.[9]

As the minister delivers his final words, an electric shock wave ripples through the room. Everyone stands and claps. Cameras flash as the applause drowns out all other sound. Stéphane Le Foll might be just an agricultural appointee of one European country but right now, for the three hundred or so people in this room, he is the Moses of the soil.

Le Foll is mobbed with photographers, handshakes, hugs, and kisses (it's France, after all). The event organizers manage to corral the various signatories onto the stage for a "family photograph." There they are—twenty men and five women representing the 4 for 1,000 nations with Le Foll at their center.

Among those in the group photo, I notice Andre Leu, an Aussie I have on my interview list. Leu is head of an organization called the International Federation of Organic Agriculture Movements (IFOAM) and as such spends a lot of time dealing with agricultural policy around the world. I ask him what the 4 for 1,000 signing event means to him. He says, "I've been going to COP since Copenhagen [2009]. And for me, Paris is the most significant. In twenty-one COPs, twenty-one years, this is the first COP that had an agriculture day. And we had the main world organizations: the Food and Agriculture Organization of the United Nations (FAO), the International Fund for Agricultural Development (IFAD), the Consultative Group for International Agricultural Research (CGIAR), and the Global Environment Fund, with thirty countries from around the world."

Leu continues in his breathless account: "All these NGOs signed on to what I believe is going to be one of the most significant changes in agriculture. Because in order to achieve 4 for 1,000, we have to change the way we do agriculture. So to me, today is the day when the world said, 'We're going to change agriculture from being a problem for climate change to being a solution.' "

Leu made another important point. While the big agricultural players (who are also the biggest polluters) had not signed on, there were over 193 countries that were represented through the Food and Agriculture Organization's commitment. And, asserts Leu, this wasn't just a piece of paper these organizations were signing. There was also a significant amount of money put forth to fund the transition. Almost $1 billion was committed by various funding entities.

Whether that money turns into real change from the ground up or is soaked up by graft and bureaucracy remains to be seen. But one thing is for sure, on December 1, 2015, Agriculture Minister Stéphane Le Foll certainly put his stake in the ground for the 4 for 1,000 program.

THE SECRET GARDEN OF STÉPHANE LE FOLL

I am once again outside the home of the French prime minister. Today the guards seem friendlier and they let me in before my liaison arrives. As I go through security I am offered a coffee.

I am shown into an office where I can sit and wait to complete my interview with the minister. This time when I am offered some breakfast and a coffee, I accept. The meal that is brought to me on fine china with real silverware is a breakfast worthy of a triple platinum restaurant in America. It's like the eggs, bread, and accoutrements have been infused with extra flavor just for the occasion of this particular breakfast. And the coffee? Well, it's as good as it gets. (I can't think of the last time I was in a US government office and was offered anything other than a stale donut, and even that had to be purchased with somebody's salary.) When I ask which restaurant the food came from my attendant is incredulous. "No," he says, "our chef prepares it here fresh. We have a full kitchen." (Of course.) I apologize and thank him for the hospitality.

Le Foll arrives to check on me just as I'm in the middle of putting a forkful of food into my mouth. I put down my fork and serviette and begin to get up and he laughs. "Please, sit. Eat. Enjoy a little bit of our food. I'll come back when you are done." I do as I am told, sitting in the antique room, looking out over a verdant garden, savoring each bite.

When my food is long whisked away by a waiter (yes, a waiter), Le Foll appears again. "Come." I follow him down a marble staircase and through the same office where his advisors were studying papers and into a beautiful garden. But now that we're here, I realize this is not a typical ornamental garden with flowers, it's a food garden.

Le Foll still walks quickly, but today he seems calmer, more meditative. "I grew up in the village of Sarthe, a small village with 256 people who worked in the rural agricultural industry. I grew up around farmers and from a very young age, I wanted to be a farmer," says Le Foll as we walk across the green grass toward a bushy area.

Behind the bushes is a huge pile of leaves and compost. Working the compost with a pitchfork is a rotund man who looks like a French tomato. He and Le Foll immediately launch into a conversation about compost. The beauty of conversations about food and farming is they translate easily between Romance languages. From English to French, "soil" is "sol," "compost" is "compost," "organic" is "biologique," and so on. It makes sense that etymologically speaking, nourishment, man's most basic need, is at the root of shared linguistics.

Le Foll explains that their garden is completely organic—no pesticides, no synthetic fertilizers, no chemical sprays for bugs, and lots and lots of compost for the soil. The tomato-shaped man is excited to show me his massive compost pile and tell me about all the things they are growing.

Le Foll continues his tour of the garden pointing out the actual tomatoes, the herbs, and the bees. He's obviously quite pleased with his little farm. I ask him if the rest of the ministers in the French government thought he was crazy to put in all these food plants. "Crazy? Eh? No, of course not. You see in France we have a very good—how do you say it? *cuisine*, so we eat what we grow in this garden."

I follow Le Foll back upstairs. We sit in the antique office. For a moment we both take in the silence as I look over my notes. "*Bon*," he says, "what else?"

Le Foll has extended me surprising hospitality. But it's time to ask him some uncomfortable questions. "Your tenure as agriculture minister will be over soon," I say. "What will happen if this 4 for 1,000 program is a complete failure? Then what?"

Le Foll takes a deep breath and lets his eyes settle on me. "If nothing happens," he says, "I fear we will have a forceful expansion of the desertification of soil. Global water and wind erosion of soil will be catastrophic. Secondly, I fear if we don't move concretely forward with this project, the biodiversity of soil itself will be lost. If we do nothing, I fear we will lose the fertility of soil. We will lose biodiversity. We will lose agricultural soil and at the end there will be escalating hunger and enormous problems in terms of feeding the planet. The longer we wait the higher the likelihood we will unleash this irreversible process. Because once it becomes infertile, it takes a very long time for soil to regain fertility."

Coming from the man who only yesterday proclaimed we could solve the world's problems with soil, this sounds cataclysmic. But herein lies Le Foll's real reason for promoting soil-based carbon capture. It is not the warming of the planet he worries about. It's feeding the planet's growing population.

"What happens when you have an agriculture that doesn't offer any more employment to a large swath of the population? It causes large migrations," explains Le Foll. "These migrations gravitate to cities. This causes more unemployment and more hunger. A state that has become destabilized is a state that has become vulnerable. It becomes the prey of those that want to take advantage of it and change it. This happened in Syria five years ago. There was a huge drought that displaced 1.5 million people and was a contributing part in the destabilization of Syria. That's what we have with Daesh in Syria and Iraq. And once that process is unleashed, the regions lose control. The agricultural goods and the costs associated with food can destabilize a country and from there destabilize a continent. Political instability is linked to hunger. We know this."

Le Foll believes what has happened with Syria is just the beginning of the type of crisis that could grow to encompass billions of people. "In the discussions we've had with African countries," he says, "they well know that they are going to see their population double in fifty, maybe forty years . . . double! If there is no agriculture there to offer employment to the youth of tomorrow, then a political catastrophe, one of geopolitical proportions, will come in its place."

For the minister there is no unlinking the climate crisis and hunger. The two are intimate results of the same root problem. I ask him how

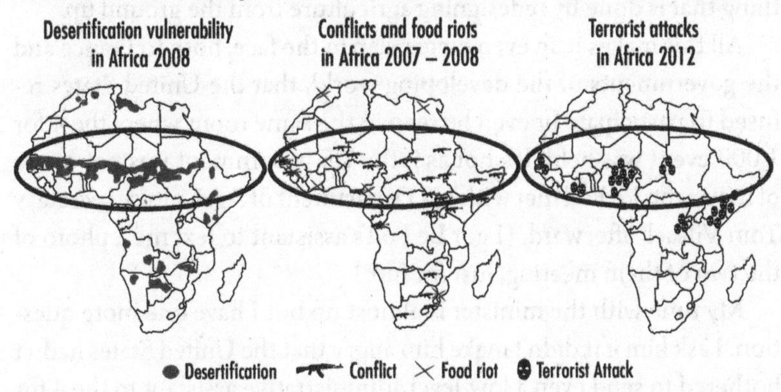

Stéphane Le Foll fears what is already happening in Africa—that degrading soils will lead to desertification, which, in turn, will lead to mass migrations and the destabilization of countries, and—indirectly—to terrorism.
Source: Global Terrorism Database.

he explains his view to people in the climate movement. "The entire world focuses on the polar bear, but there's another reality," he says. "If we let climate change get out of control, there will be millions of climate refugees. Millions! Where will their populations go?" As countless refugees currently pressure European countries for new homes, Le Foll's point is clear. Shocking footage from the French border cities shows refugees pouring over fences like rivers of human bodies. These disturbing real-life videos look like scenes from the all-hope-is-lost Brad Pitt zombie movie *World War Z*.

"That is why everything we are developing in France for agroeconomics is to change this cycle. To ensure the transformation into what we call agroecology or intensive ecology," Le Foll explains. "To intensify the natural mechanisms in agriculture and not increase the use of machines or chemicals because we are at the end of that process. This is the only way we can feed the world. I am absolutely certain of it."

When we talk about feeding the world in the United States what we really mean is increasing profits for agricultural and food companies. But when Le Foll speaks about feeding the world, he is speaking directly about feeding impoverished countries in places like Africa. And as he sees it, feeding the world is not something you do by giving handouts of overproduced grain made through chemicals, it is something that is done by redesigning agriculture from the ground up.

All this makes it an even bigger slap in the face, both to France and the governments of the developing world, that the United States refused to participate or even be seen in the same room where the 4 for 1,000 event was held. It's not as if the US government was not aware of the event. Le Foll met with US Department of Agriculture secretary Tom Vilsack afterward. (I got Le Foll's assistant to text me a photo of the two of them meeting, just for fun.)

My time with the minister is almost up but I have one more question. I ask him if it didn't make him angry that the United States hadn't bothered to send even a low-level administrative assistant to the 4 for 1,000 signing. His answer is careful and diplomatic. "We are at the beginning, I hope, of a beautiful project that perfectly joins the fight against climate change with the fight against world hunger. Yes, it's true that the US didn't sign it but we are generally in agreement that carbon storage in soil is a common goal. Why did they retreat from it? Because in American agriculture there exist huge goals that don't concern the soil, they concern the industry."

I tell him he hasn't answered my question. I want to know if it personally offended him that the United States did not show up. He dodges me again: "At the very least, we must try to cooperate. Okay, at COP21 the US brought their project called GACSA." GACSA, again, is the Global Alliance for Climate-Smart Agriculture, the Monsanto-backed nonprofit that promotes genetic engineering and to which the United States and other countries have signed on. Le Foll explains that "this climate-smart agriculture does not permit nongovernmental organizations from truly participating in their agreement."

This is because all the technology for genetically engineered crops is patented and owned by a few corporations. NGOs can't honestly participate because there is no room for the "teach a man to fish" model here. It's all based on control of the source of food (the seed) and the resulting profits.

Le Foll continues, "It is not a project on a truly global scale. It isn't dynamic enough to carry the weight or need of what we have here." He won't come out and say it but it is obvious that climate-smart agriculture is a means to sell and profit from patented technology.

Behind the lights and cameras and very much behind the scenes there was a showdown at COP21. It was a showdown between the 4 for 1,000 initiative and the GACSA-proposed, genetically engineered "Climate Smart Agriculture." While there will be no shoot-out at the OK Corral per se, there's plenty of pressure from those who would prefer to see Le Foll, biosequestration, 4 for 1,000, and agroecology go the way of the dodo bird.

I press Le Foll again, asking him if it doesn't just chap his hide that the largest agricultural producers in the world basically snubbed him. "Yes, it's . . . frustrating . . . yes, well, it is frustrating!" he admits in a rare burst of, well, frustration. He quickly recomposes himself. "*Bon.* We can't leave the US out. We will continue to try to make the US stand with us. In the end I hope they will sign."

What he knows and is not saying is that without the United States, the 4 for 1,000 program is missing one of the greatest single agricultural landscapes capable of sucking up carbon.

It's true that US farmland only occupies a fraction of the 5 billion hectares of land under grazing and cultivation on the planet, but because of the type of intensive cultivation that can be done in the wide-open spaces of North America, the land has a disproportionally high carbon sequestration potential. For 4 for 1,000 to truly work on a global level, the United States will have to participate. This is a daunting task, considering the politics of profit at play.

Not to be left without an uplifting end moment, Le Foll gives me one more zinger. "I've always kept in my mind this vision: the 60 centimeters of soil everywhere on the planet, it gives us life, allows us to have livestock, to produce fruits, vegetables, grain, milk . . . but the Earth's soil, we have forgotten about it. We walk on it, step on it, we don't even notice it's there. Soil needs to be placed at the center, at the heart of this debate," he says, concluding our interview.

WAG THE POODLE

It's Saturday morning, December 12, 2015, and I'm back in California. The newsfeed from the other side of the world in Paris shows a throng of well-dressed people in the COP21 convention hall. There in the front row at the big table are UN secretary-general Ban Ki-moon; the French foreign minister and president of COP21, Laurent Fabius; and the French president (whose car nearly ran me over in the courtyard of the Hôtel Matignon).

It's time to announce the Paris Agreement. Ban Ki-moon joyfully declares, "Today, we can look into the eyes of our children and grandchildren and we can finally say, tell them that we have joined hands to bequeath a more habitable world to them and to future generations." French president François Hollande says, "December [twelfth] 2015 can be not only a historic day but a great day for humanity."

As the big moment draws near, the Twitter feeds for #cop21 and

#climate light up. Then finally, it happens—the little green gavel slams against the table and the agreement has been formalized. The camera cuts to Al Gore with a huge TV-smile, clapping like somebody from the Tennessee Titans just made a touchdown. A little man to his right hugs him awkwardly (Mr. Gore doesn't hug back). As if on cue, the feed cuts to a shot of two women crying. Tears of joy stream down their faces as they embrace. We cut back to an ocean of diplomats clapping.

People are breathlessly posting on social media about this "historic moment." Twitter goes ping again. There's astronaut Scott Kelly tweeting from aboard the International Space Station, "#Congratulations #COP21 Delegates on your historic agreement! #Earth thanks you and so do I! #YearInSpace."

To add a little flare for the cameras, the leaders at the big table stand and hold hands high above their heads. It's sort of like a football game–inspired crowd wave to say, "Go World! Go Humans! We did it!" Then a couple of the women from the big table hug. It's all so . . . well timed, well lit, uplifting, and picture perfect. No Hollywood studio could have done a better job.

After a little while President Obama appears on TV. He's back in the United States too. "A few hours ago, we succeeded. We came together around the strong agreement the world needed," he says. It's a perfect end to the perfect story of international goodwill and brotherly love. Like other people in the free world, I can now shut my laptop, turn off my TV, and go about my weekend feeling good that our fearless leaders have delivered humanity and the planet unto safety.

But a closer look at the newsfeeds shows that not everyone at COP is smiling and clapping. At least one camera catches the look on the faces of seventy-year-old Marshall Islands minister of foreign affairs Tony de Brum and his two associates as the gavel comes down. Their expressions are a mixture of shock and sorrow. Even with a group of eloquent youths from the Marshall Islands getting tremendous screen time during the convention, the Paris Agreement does nothing to halt the sea level rise they fear will soon devour their homes. Sorry, Mr. de Brum, but as the *New York Times* put it, "The Marshall Islands Are Disappearing."

De Brum and others who are being directly impacted by the changing climate of our planet may have been upset, not because they are party poopers, but rather because of what was actually in, or rather not in, the final Paris Agreement that was being globally celebrated. The nature of the agreement is largely already spelled out (in most cases word for word) in the 2009 Copenhagen Agreement, which apparently ended in disagreement. Thus, the Paris Agreement is really an agreement to finally agree on the terms set forth six years earlier at COP15.

So what were those terms?

Essentially, in yet another five years (in 2020), the nations of the world agree that they will initialize plans to limit carbon emissions. From that point on, they plan to "ratchet" down their annual emissions. In other words, after twenty-one years and twenty-one COPs, the big moment in Paris was an agreement to wait yet another five years to begin to limit emissions. (In terms of diplomatic job security this is good news, as there will be many more COPs.) The idea is that these limits on emissions will ensure the average increase in global temperature will be "well below 2 degrees Centigrade" (3.6 degrees Fahrenheit).

But assessments of the 180 national climate-action plans submitted by countries at the summit suggest that even if these plans are followed, the world will see a temperature rise of closer to 3 degrees Celsius (5.4 degrees Fahrenheit) by the end of the century. These plans are aptly called "Intended Nationally Determined Contributions" (INDCs), because they are completely voluntary. There is nothing to legally bind the countries to achieve their proposed numbers. If history is any guide, these good intentions will play out with some nations doing a lot while others put short-term economic profits ahead of reducing their carbon dioxide emissions.[10]

Inside the COP21 agreement, there's also no accountability in reporting carbon dioxide emissions, and not surprisingly, very little transparency. No process even exists to verify the emissions of the 195 participating countries. INDCs are presented and calculated in a smattering of formats because there is no standardized format for accounting and reporting.

After twenty-one years, it is a surprising oversight that nobody created a standard Microsoft Excel spreadsheet template that participating countries must fill in with their standard carbon dioxide emissions, accounted for in a standard way. (Bill Gates himself was actually at COP21 but not even he created such a spreadsheet.) Such a glaring omission makes one wonder what really goes on behind closed doors at these conferences.

One also wonders what type of leaders are sent to negotiate such agreements. Certainly missing are those with a basic ecological education. As such, the Paris Agreement fails to address the two largest "sinks" for carbon: the oceans and the soil. It appears that in two decades of climate talks the fearless leaders of Planet Earth, put in charge of reducing carbon dioxide emissions, have not been given a high school–level course on where carbon is stored and why.

The Paris Agreement avoids at all costs the word *pledge* and the word *commitment*. In a last-minute move before the signing, representatives from the United States insisted on replacing the word *shall* with *should* in relation to industrialized nations' responsibility to reduce their own emissions. There are no target numbers for countries to achieve. Any language about mechanisms for achieving reductions (by ramping up solar, etc.) was left intentionally vague.[11]

For all the hoopla about fossil fuels ending life as we know it, the Paris Agreement also contains no wording about "decarbonizing," no wording about "reducing fossil fuels," no wording about "hydraulic fracturing" or "tar sands" or "gasoline" or "coal." After twenty-one years, it appears the collective leaders of the world are incapable of writing these words in conjunction with an accord having to do with reducing emissions from these very sources. This flies in the face of the Intergovernmental Panel on Climate Change (IPCC), which was founded by UN charter and which points to burning fossil fuels as the leading cause of the rising atmospheric carbon dioxide emissions.

The result of no transparency, no legally binding agreement, no ecological methodology, voluntary emissions reductions, and no clear language around solutions to rising atmospheric CO_2 is that, like COP21 itself, the Paris Agreement is pure hyperbole. Like the other

COPs, this was political theater. From the outset it was expertly choreographed to distract from the reality that apart from some carbon emission reducing concessions by a few small and well-meaning nations, there is absolutely no viable plan to deal with rising atmospheric CO_2 emissions.

Not surprisingly, Lord Deben and Lord Krebs of the Committee on Climate Change report that "a number of leaders have returned to capitals, Paris Agreement in hand, only to announce that domestic policy need not change in light of the new global deal."[12] Apparently some of those leaders are from countries that had lavish displays at COP21, countries like India.

India's current plan to provide electrification to its over 1 billion citizens involves building 455 new coal power plants. That is on top of the 148 coal plants it already operates, which are among the dirtiest and most inefficient in the world. Prime Minister Narendra Modi, who ran under the banner of solar energy, has since denounced climate change science in favor of a rural electrification plan that involves ramping up production from some of the largest coalfields on Earth.

However, Modi has offered a way for India to go solar. He says that India will go green so long as the developed nations foot the bill. The cost? A cool $2.5 trillion. At over $166 billion a year for the next fifteen years, that's quite a chunk of change for the world to pay for India's solarfication.[13]

As of yet, no nation or group has stepped up to pay for India's green electricity, and India's plans to go gonzo for coal are full steam ahead. Should India turn on its coal-fired power plants, all the carbon saving that all the nations of the world have done thus far will be moot.

Perhaps this is why, when Mr. de Brum of the Marshall Islands made pleas for his small sinking island nation at COP21, Indian environment minister Prakash Javadekar responded by saying, "So what?"[14] After all, the industrialized nations of the world built their strength with fossil energy and now it's India's turn.

When I ask author and environmentalist Paul Hawken about his reaction to COP21, he tells me something that rings true. "I think there's a strong desire to have a kind of sort of charismatic event that tells us that we've passed from one regime to another, and COP21 was that event," he explains. For years, Paul has been working with a group of scientists on a list of one hundred solutions to carbon dioxide emissions. He calls the project and its accompanying eponymous book "Drawdown."

Paul says that moving swiftly to renewable energy and energy efficiency are important, but neither will do anything to draw down the legacy load of carbon that is in the atmosphere. "If we want to achieve drawdown," says Paul, "we have to go thank the Earth, pay homage to it, and start to farm and grow our plants and trees in a highly different way than we have for the last one hundred fifty years."

Indeed, this is the same message that Stéphane Le Foll and his team are trying to get world leaders everywhere to sign on to.

Call it regenerative agriculture, agroecology, biosequestration, or drawdown, the United Nations and most of the world are currently blind to the one simple solution to carbon emissions. The irony is that bringing carbon into the soil solves multiple global problems. It reduces carbon dioxide in the atmosphere, it increases the fertility of soil, it helps farmers grow more, and it allows the oceans to release the CO_2 that threatens to acidify the phytoplankton that produce so much of the oxygen we breathe. (As the concentration of CO_2 in the atmosphere is reduced, the oceans naturally release excess CO_2 back to the atmosphere.) In essence, it doesn't matter which of these issues you care or don't care about. It only matters that we align on the single overarching solution in front of us.

If Le Foll, Hawken, and others are right, then what we eat has a far greater impact on the future of life on Earth than what we drive and how we get our electricity. The problem, at least in America, is that what we eat comes from a large and complex agricultural system. And for a long time that system has had much more to do with death than with life.

CHAPTER 2

NAZIS AND NITROGEN

I am standing next to a large steel green and yellow combine in the middle of an ocean of "field corn," the type of corn grown to feed animals and the most common crop in North America. The combine isn't special. In fact, its guts were developed long ago, during World War II. There's nothing particular about this cornfield either. In some ways its uniformity, its blandness, is calming. It's like being inside a yellowish ocean. But it too is a result of the global conflicts that consumed much of humanity's attention during the first half of the twentieth century.

Riding inside the combine with my host, David, a tall, strapping man whose Scandinavian grandfather farmed this land, makes me want for the comfort of my rental vehicle. Yes, there is modernity—a specialized farming GPS has been crudely affixed to the windshield and a radio tuned to an AM station loudly babbles about local politics. But no, this is not the air-conditioned luxury we city dwellers imagine most farmers riding around in.

The seat is uncomfortable. The air is filled with a light haze of corn grit. It's so loud that David and I are both wearing ear protection (hence the radio being at full blast). And the novelty of watching the teeth of this machine mow down endless rows of corn wore off an hour ago. It's like being cramped in the backseat of an old school bus going over a bumpy dirt road, forever.

Like the vast majority of farmers I meet, David and his family have a modest house. Its single-story brick exterior is so square, so neat, that it mirrors the cornfield across the road. Nothing in its interior, including the amount of space, is lavish.

I ask David if I can take a photograph of him and his family outside their home. His wife and three children are handsome and wholesome-looking. They stand, self-consciously, on the neatly trimmed lawn in front of their small dwelling. They say "cheese." The American flag flutters as if on cue behind them. *Snap*. There it is, the picture-perfect modern American farm family.

If I walk away now, without asking any questions, I can rest assured that all is well in America's heartland. Sure, Old MacDonald has a bigger, more uniform operation than one might expect. But so what?

The wind is picking up, and it's getting cold. I think about going back to the hotel, getting in bed, eating those cookies I know they'll have waiting for me in the lobby and watching a couple of *Seinfeld* reruns.

Then, against my better judgment, I ask David if he can spare a few more minutes to show me around.

Like most of the American farmers I meet, Dave and his family are genuinely nice people. They are not a loquacious bunch, but if you're stuck on the side of the road with a flat tire, they're the first people to stop and help.

There's something else about these "heartland" farmers, something more difficult to describe. It's a combination of sense of duty, honor, love of what they do, and a deep understanding that no matter how badly they are treated by the media, no matter how much they are looked down on, we need them.

Of course, David says, he's happy to show me around as he gestures toward his steel barns.

As we walk I ask him why he only farms corn. He explains that his grandfather had farmed a diverse array of crops and animals, but as time has gone on there's been more demand for corn and soy. I ask him if he knows what his corn goes into. Not only does he know, but he recites the percentages. About 43 percent will go into animal feed, 28 percent will go into ethanol fuel, 13 percent will be exported, 11 percent will go into processed food and beverages (mostly as corn syrup), and 5 percent will go into industrial chemicals and things like biodegradable plastic cups.

David says he struggles with the fact that those numbers don't exactly reflect that he and other corn "producers" are "feeding the world," as the slogan goes. But in many ways corn does form the basis for today's American food pyramid. For instance, says David, the type of corn he and most corn farmers grow is to feed the animals Americans eat. Hence the common misnomer "feed corn." As Michael Pollan aptly points out in *The Omnivore's Dilemma*, corn is such a foundational part of our food supply that it now leaves chemical markers in our hair.

The catch-22 with moving to corn, David says, is as farmers like him get more efficient, they either have to find more ways to monetize their crops or they go out of business. Standing in the barn looking at the array of mega-sized Tonka trucks that can till, spray, harvest, and carry grain, I ask David what he means by "efficient."

He explains that the per acre yield of corn has skyrocketed since his grandfather's day. His granddad was lucky to get around thirty or forty bushels per acre. In contrast, today in the noisy combine "we" harvested around 150 bushels per acre—and some of the farmers he knows are pulling in up to 180.

When asked about the inputs and investment needed to squeeze that kind of productivity from the land, David says they've all gone up too. Farm chemistry can get complicated but the basic rules of application, he says, are simple. The more dry weight of corn (or soybeans) you want, the more pounds of nitrogen (N), phosphorus (P), and potassium (K) you add. But adding more inputs only works up to a point, which is called your "maximum yield" (MY). And gauging exactly where that point is so you don't spend unnecessary money on inputs? Well, says David, that's somewhere between a "scientific guess" and a "a lot of prayin.'"

He explains that the basic numbers for application break down as follows. To grow an acre of corn today you apply around 140 pounds of ammonium nitrate (nitrogen), around sixty pounds of phosphate (phosphorus), and around eighty pounds of potash (potassium).

Added to that are about two to three pounds per acre of herbicides (like glyphosate, the primary chemical in Roundup), insecticides, and/or fungicides.

For David's farm, a two-thousand-acre postage stamp in the middle of America, these numbers quickly add up. His investment in seed, chemicals, and equipment each year runs into the millions. If his crop fails, government insurance will cover a minimum price for what he's planted. But that doesn't stop him from waking up at 4 AM every day and working until well past sundown.

David's family income has been better in the past several years, he says, due to the "bump" that ethanol fuel has given him and other corn farmers. It's the only buffer against the rise and fall of the market price of corn (which is set by futures traders in the stock market). That extra bit of cash from alcohol fuel has allowed his wife to ease up on her off-farm job hours. He fears the added security won't last so he and his wife are trying to start a side hunting-tour business. She has kept her job and still works part-time to make ends meet.

THE FIELDS OF SISYPHUS

In almost all the farmers I meet there's a pervasive sense of fear. The terror of "losing the farm" is what drives many of them to toil endlessly. Most are aware they are caught in a Sisyphean trap. The more they produce, the lower the market price for their commodity, which means the more they have to produce to stay in business because their margins are forever shrinking. (Push rock up hill, let rock slide down, repeat.)

Efficiency (also known as "overproduction") is the thing that has pushed millions of farmers off their land. At the height of farming in the United States in the 1930s, there were around 6.5 million farms. Today that number stands closer to 1 million. (The USDA puts the number closer to 2 million, but that includes about 1 million tiny hobby farms.) The number of farms is still shrinking. In fact, America loses about two farms every hour, twenty-four hours a day, 365 days a year.

The vast majority of the surviving farms are husband-wife enterprises with the husband listed as "the farmer." Those 1 million farmers produce the food that the other 324 million of us eat. In other words, the great efficiency of modern farming now makes it possible for every 1 farmer to feed 317 nonfarmers. It's really a miracle.

More food, more cheaply from fewer acres has provided the basis for unprecedented global population growth. At the birth of the modern agricultural petroleum-fueled revolution in 1927, humanity stood at a mere 2 billion people. As I write these words, our species numbers some 7.3 billion people. According to the United Nations, we will add yet another 2.4 billion humans to the planet by 2050 for a total of 9.7 billion by the middle of this century.

So much food. So many humans. Yet another miracle.

But these miracles—modern farming's ability to produce nearly infinite quantities of calories and the global population boom—are on a collision course of unprecedented proportions. Both miracles come at a great cost. The chemicals needed for single-crop (aka "monocrop") farming are titanic in quantity, nonrenewable, and toxic. Meanwhile, the increasing number of humans, with their near-insatiable appetite for commodity crops and derivative products, adds ever more pressure to pump out more calories, which necessitates the use of exponentially more toxins. It is Sisyphus writ large, a cycle born from conflict that can only end in suffering.

Food and war. They seem so diametrically opposed, so intrinsically incompatible. But historically, the better human beings became at war, the more facile we became at applying the technology of carnage to mass-produce food. It is thus no accident that the way we produce food today is, by design, a war with nature.

They say that history repeats itself and the history of agriculture is no exception. Civilizations grow until they either collapse or find yet another finite resource with which to temporarily prop up their soils. For the first time in history we are now playing this game at a global scale. Formed by desperation and forged by war, today's agriculture is so vastly altering our landscapes and so quickly outpacing the Earth's ability to regenerate soil that agriculture now threatens planetary life itself.

The last century of agricultural history tells us what went wrong with our food system and why. But more importantly, this history may contain the secrets to unlocking a way to save our food and ourselves.

UNDER PRESSURE

At the start of the nineteenth century, Europe and America rapidly industrialized and populations in cities quickly expanded, creating a spike in demand for that ever storable transportable foodstuff: grain. Farmers on both sides of the Atlantic began looking for new sources of nitrogen with which to increase crop yields. By then it was widely accepted that to grow large quantities of a crop one needs a balanced mixture of three elements: nitrogen, phosphorus, and potassium. While phosphorus and potassium could be mined, getting nitrogen was more difficult. For a time, bird guano from South America provided the nitrogen needed to fuel the West's agricultural expansion. But demand quickly outstripped supply.

The problem with nitrogen is that aside from guano, saltpeter, and the manure of livestock, the vast majority of nitrogen on Earth is locked in our atmosphere. Air is about 80 percent nitrogen.

Chemists around the world were trying to figure out how to get nitrogen out of the air and into a form that could be readily applied to soil. A Jewish chemist

While there are numerous natural processes that bring nitrogen from the atmosphere down into the soil (including appropriate livestock grazing and cover crops), modern agriculture is based solely on synthetic nitrogen.

working for the BASF chemical company in Germany by the name of Fritz Haber was the one who cracked the code. In between his hobbies of romantic poetry and fencing, Haber had been thinking about how to synthesize nitrogen.

Haber decided to try a process called hydrogenation, which involves heating hydrogen under pressure. His laboratory forays were soon rewarded with a few drops of ammonium nitrate. This was enough to convince another man working for the BASF company, Carl Bosch, to fight for Haber's idea.

Bosch worked with Haber to better his design. He then spearheaded a multiyear campaign to fund and build a state-of-the-art processing plant capable of producing nitrogen synthetically. In 1913 a towering new facility capable of high-pressure hydrogenation was turned on just a few miles from BASF's headquarters. The facility was soon producing hundreds of tons of ammonium nitrate for use as fertilizer.

The implications of the Haber–Bosch process are so vast and far-reaching that some historians believe this one industrial achievement underpinned the entire multibillion-person increase in human

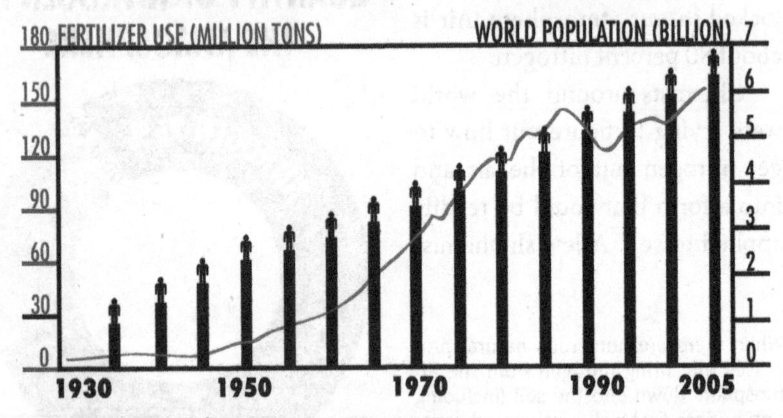

Would the global population have expanded as quickly without the advent of synthetic nitrogen?

population that has occurred since 1913. But this one invention also played a pivotal role in a much darker chapter in human history.

BETTER LIVING THROUGH CHEMISTRY

Only months after the first Haber–Bosch plant was turned on, Germany barreled headlong into the First World War. The German government enlisted Haber and Bosch to assist the military in its war effort. At first the two were commissioned to build a new facility to make nitrogen for bombs. But when the new plant took too long to build the two chemists focused their energy on creating a different kind of weapon.

As soon as Haber's synthetic nitrogen was sprayed on crop fields it had an unintended side effect; synthetic nitrogen didn't just make crops grow big and strong, it also caused an explosion of weeds. Added to the weeds was yet another unforeseen problem: pests. For the first time, synthetic nitrogen gave farmers the power to grow vast acres of a single crop. But a field of the same crop attracts many more pests than a diverse field. To combat the weeds and pests, Haber concocted new poisonous chemical compounds that would come to be known as insecticides and pesticides.

Through his work with the German military, Haber quickly found new uses for his lethal chemicals. Haber personally designed and oversaw the first-ever use of poisonous gas as a chemical weapon. As the war ramped up and the Allies also gained chemical weapons, Haber continued to expand his repertoire of chemical kill agents.

Haber's first wife, also a brilliant chemist, was so disgusted with his oversight of Germany's chemical weapons that late one night she walked outside their home and shot herself in the chest with his service revolver. In spite of her suicide, the next morning Fritz Haber left to oversee a chemical weapons attack against the Russians on the Eastern Front.

Germany's bravado and lack of foresight would soon find it on the losing side of that war. In the aftermath of reparations to the rest of Europe, Germany consolidated its big chemical companies under one umbrella. The new enterprise was called IG Farben. And it would be headed by Haber's partner, Carl Bosch.

As the Nazi Party rose to power and Adolf Hitler took center stage, Bosch and IG Farben entered into lucrative long-term contracts with the Third Reich. On the other hand, Fritz Haber and other high-ranking Jewish people who were lucky enough to survive fled Germany.

As Carl Bosch would learn, one should not make a deal with the devil if you do not wish to become one of his demons.

THE DEVIL'S WORKSHOP

Perhaps no single event has had a deeper or longer-lasting impact on how we eat than the Second World War. One could argue that all the elements that would come to make up the modern food system were already in play, from synthetic nitrogen to chemical weapons, to price-fixing corporations to the cycle of farm overproduction. But elements by themselves do not make a system. To make our modern food system, it took a global war.

The First World War had been fought largely in trenches, but this new war demanded a level of centralized control, mechanization, petroleum power, and efficiency. Together, these disparate pieces formed a new whole. The purpose of this new sociopolitical-industrial war machine was the biggest bang for the buck. Nowhere was that more pronounced than at the world's largest chemical company.

At the 1936 Summer Olympics in Germany, IG Farben's good corporate citizen face was plastered everywhere, with ads for its Bayer aspirin and posters for its floor polish, saccharine, soap, and other products. But the IG Farben name was superseded by a different kind of sign that was also plastered across Berlin: the Nazi swastika. It was the union of these two entities, the Third Reich and the largest chemical company the world had ever known, that gave the Nazi war machine its edge.

By the time the Olympic torch was paraded in front of a sea of Nazi swastika banners, IG Farben's profits were already coming largely from its sale of goods to the Nazi military. In the three years leading up to 1939, IG's profits leapt 71 percent as its 230,000 employees busily prepared for the biggest war the world has ever seen.

Using the hydrogenation technology originally perfected by Haber and Bosch for nitrogen fertilizer, IG Farben was busily creating synthetic fats used for food. Its synthetics also included gasoline, rubber, and nitrates. The corporation's stamp was on a long line of explosives like TNT, hexogen, and nitropenta, which together formed about 85 percent of Germany's firepower. The company was responsible for about 25 percent of the products that made up each Nazi soldier's equipment, right down to his boots. It made metal for tanks, nickel for engines, aluminum for bombers, lubrication for engines, armor for vehicles, and even the plastic for the keys for the infamous Nazi encrypting machines. But all that pales in comparison to the twisted nature of the company's most lethal wartime inventions.

Using the herbicide, insecticide, and chemical weapon technology developed by Haber, IG Farben weaponized mustard gas and concocted chemicals like tabun gas, which attacks the central nervous system, leading to fatal contractions. The development of tabun led to the invention of sarin gas (isopropyl methyl phosphorofluoridate), the smallest inhalation of which brings about a horrific and painful death.

But by far the darkest of IG Farben's exploits were in the concentration camps. The cartel used the labor of prisoners to construct a massive hydrogenation plant capable of producing both synthetic fuel and rubber. Adjacent to that chemical plant the company built its own concentration camp. They called their new chemical facility IG Auschwitz.

Soon thereafter, IG Farben used Haber's chemical weapons to their ultimate potential. In concentration-camp gas chambers across Germany, a compound named Zyklon B, a deadly gas that is a combination of hydrogen cyanide mixed with an adsorbent, was used to kill at least 1 million Jews. Some of those who were gassed to death were relatives of Fritz Haber, himself a Jew.

IG Farben also funded and carefully monitored the experiments of Dr. Josef Mengele. Working with scientists from the IG cartel, doctors from the SS, and officials from the Nazi Party, Mengele dreamed up a long list of chemical and biological weapons that he injected into humans, killing them in the most horrific ways imaginable. Along with similar IG Farben–sanctioned experiments at other concentration camps, Mengele's experiments led to the development of new pharmaceuticals as well as advanced chemical gases, liquids, neurotoxins, and biological weapons.

In the aftermath of the war, as much IG Farben intellectual property as could be found was taken to Great Britain and the United States. Laboratories, documents, and pieces of equipment were shipped out of Germany for further study. Meanwhile, big American firms cherry-picked IG Farben's top engineers and chemists by offering them lucrative salaries and new lives in the "land of the free."

With little in the way of reprimands and a wealth of new knowledge of the power of chemistry to shape the world, IG Farben's chemicals, processes, techniques, and secrets would soon find their way onto American store shelves. From plastics to manufactured goods to drugs and most notably to food, much of the industrial West's post–World War II boom would bear the invisible fingerprints of the Nazi-run chemical giant.

With its patents moving internationally from corporation to corporation and many of its engineers resettling in the United States, IG Farben's deadliest chemicals were also repurposed, right down to the widespread use of Zyklon B as an insecticide on American farm fields.[1]

MEANWHILE, BACK AT THE RANCH

At the close of World War I, European agriculture was in disarray. American farmers soon came to the rescue. "Plow to the Fence for National Defense!" and "If You Can't Fight, Farm" were the slogans that put a jaw-dropping 40 million new American acres into food production. The result was a brief spike in farm income that rose sharply from an estimated $5 billion in 1916 to roughly $9 billion by 1919.[2]

Heeding signals from both market and government that more food was desperately needed, American farmers borrowed heavily against their land to purchase yet more land, more seed, and more equipment. This spending spike lured unscrupulous land speculators, who inflated the prices of farmland and, in so doing, added to the ballooning national farm debt. Meanwhile, Europe's farms regained their footing and demand for food exports from America's breadbasket suddenly plummeted.

By 1921, just two years after its peak, American farm income crashed to $3.4 billion, its lowest so far that century. Seeing high food prices, farmers had planted more seed in more land with more equipment, which in turn led to overproduction of foodstuffs, which then led to too much grain and finally to depressed prices. The uncertainty of the years that followed the farm crash was only made worse by the onset of the Great Depression. Farm income fell an additional 52 percent from 1929 to 1932.[3]

Thus began the cycle of boom, overproduction, and bust that still plagues American farmers today. This "more is better" mentality still drives modern agriculture with its pedal to the metal to squeeze every possible calorie from every available acre of land. Most farmers realize this cycle is self-destructive. But politicians and agribusiness companies keep singing the same song: that it's American farmers who can, should, and must "feed the world."

Even before the Great Depression farmers had begun to focus their grievances toward America's East Coast and the seat of political power. The Grange, the National Farmers Union, and the Farmers' Alliance were among a group of organizations that coalesced into a renewed progressive movement arising from the Midwest. In an effort to stabilize grain prices, this "farm bloc" called for abolition of futures trading, the practice of selling and buying grain before it is produced.

Famers wanted to set their own prices. But the traders in New York City were not having it.

As farmers organized, fear of a populist revolt was growing in Washington, DC, and New York City. When President Franklin Delano Roosevelt (FDR) entered office in 1933 he immediately created a program to pay farmers not to grow. Millions of acres of cotton and millions of hogs were destroyed, all in an attempt to curtail runaway overproduction and stave off the downward price spiral of farm commodities.

FDR knew he had to maintain peace among his powerful East Coast supporters, but he also knew that farmers had a point. As long as a handful of people dictated the price of grain, farmers would always be pushed to grow more and accept lower prices. Like his distant relative President Theodore Roosevelt, FDR believed that the American economy is built on a free market and he had also run on the promise of "trust busting," or breaking up the large monopolies.

President Roosevelt espoused that this open marketplace system inspires diversity, competition, and resilience. Conversely, he believed an industry that controls an entire vertical (food, for instance) would have undue economic power (to fix its buying prices low and its selling prices high) and would certainly wield undue political power (to influence legislation). But try as he might, FDR and subsequent US heads of state would learn that it is extremely difficult to reestablish a free market once it has been infected by the disease of monopoly.

By the time President Roosevelt attempted to curtail runaway farm production and install mechanisms to properly pay farmers, centralized control and distribution of food had clenched a fist around America's breadbasket. The railroads and grain traders were firmly controlling both the prices and the flow of grain through the Midwest. Without the rails, farmers had no way to move their product to market. Without the traders in New York City, there was no market.

Prices were set far away on Wall Street and posted at rail spurs across the Midwest. A farmer had two options: sell his grain at the posted price or go home with no money in his pocket. Despite all of FDR's attempts at antitrust and farm policy, this centralized pricing system still exists today.

Aside from farmers' markets, the price of farm goods is set not by farmers but rather by the US government, commodity traders, consolidators, packers, and distributors. It is these entities, not the farmers, who are collectively referred to as "the world food market."

GET YOUR FARM IN THE FIGHT!

During World War II midwestern farmers faced a new problem. With farm labor taken by the factories in the cities and the battlefields overseas, there were not enough people to farm. With pressure to produce more food for the war and less hands to do it there was only one thing to do. Farmers had to become more efficient. And to do that they needed new toys.

"*Here's the Tractor* THAT REALLY HELPS WOMEN AND CHILDREN TO DO STRONG MEN'S WORK!" proclaims a 1943 Ford-Ferguson tractor advertisement. The small print in the ad emphasizes, "The Ford Tractor with Ferguson System and Ferguson Implements make it possible for women, children, and old people to swell the ranks of farm labor and give valuable aid in meeting the tremendous food production goals of the future." The ad is not exactly a Madison Avenue award winner, but for women left to manage the family farm, it hit home.

Bolstered by the demand for tanks, guns, and airplanes, American agricultural machine companies received the equivalent of a steroid injection. Massey-Harris churned out tanks, aircraft wings, and trucks. John Deere built tank transmissions, aircraft parts, and ammunition. Case made bomber wings, aircraft engine parts, and artillery shells. Allis-Chalmers machined turbines and propeller shafts for ships. That company even had the honor of building the casings for "Little Boy" and "Fat Man," the two atomic bombs that were dropped on Japan.

Manufacturers used their new technology to build new types of agricultural machines. Hydraulics, power take-offs, and three-point hitches, all wartime innovations ported from the battlefield, became standard on farm tractors. Meanwhile, tractors got smaller, easier to manage, and much more powerful.

What's more, tractors weren't just pulling or pushing implements, they were gaining the ability to be hooked up to complicated "transformers" that could perform many tasks. Massey-Harris applied its wartime engineering lessons by building the first self-propelled machine that could cut wheat and separate it from the chaff. By *combining* the two labor-intensive tasks central to the primary grain of civilization, the machine represented a watershed moment in agricultural technology. Enter "the combine."

When Massey-Harris wanted to produce more self-propelled combines than wartime quotas allowed, it convinced the War Production Board to allow it to do so if each buyer signed an agreement to produce two thousand acres of wheat. They announced their program in newspaper ads that loudly stated: "WANTED! Farm front warriors for the 1945 HARVEST BRIGADE." *Fortune* magazine did an eight-page spread on the "effort," and Massey-Harris paid for a twenty-minute Technicolor film to advertise it. Not surprisingly combines quickly went on backorder. Tractor sales alone jumped from 1.6 million in 1940 to 2.4 million in 1945.

Thanks in no small part to the largest war in human history the age of mechanized agriculture was born. The caloric output of America's farm fields soon soared and farm income almost tripled from $4.4 billion in 1940 to $12.3 billion in 1945 at the close of the war.

But all this was just a warm-up for the new war that was soon to come.

THE FOOD BOOM

By the time the Allied forces hit Normandy, France, in June 1944, Germany's knowledge of chemicals and America's facility with machines had reached unprecedented heights. Men would soon return from the battlefront. New marriages would occur. A global boom in babies would produce the largest generation that had heretofore existed.

To feed that new population a different kind of marriage was also consummated. This new union fused Germany's deadly chemicals with America's mighty machines to produce the largest agricultural powerhouse in the history of civilization: the chemical industrial–agricultural complex.

Just as the Germans had learned after they adopted synthetic nitrogen, American farmers were finding barriers to producing as much wheat as possible on land that heretofore had a rotation of different crops and animals. The first barrier to unlimited acres of the same crop was pests, the second weeds, and the third fungus. Without balanced soils, which have inside them all the microbial life needed to support plants, nature will cull a crop. In nature, diversity is the norm, not the exception. So an ecosystem in a state of unbalance (too much of the same plant) will, through bugs, weeds, plant disease, et cetera, attempt to restore itself to balance (diversity).

The end of the war brought with it relief for Midwest farmers on the calorie-producing treadmill. Wartime manufacturing restrictions were finally lifted. Shiny new agricultural machines were available at dealerships nationwide. And importantly, for farmers struggling with crop blight, army ants, grasshoppers, corn borers, and a horde of other bugs, there was suddenly a plethora of new sprays and powders that promised to magically eliminate these pesky problems.

The pesticide industry used the wartime technology of chemical weapons manufacturing to initiate a massive expansion, and $3.8 billion was invested in new infrastructure and facilities to produce chemicals for US agriculture.[4] In the first seven years after the war some ten thousand new pesticides were registered with the USDA. Not surprisingly, chemical industry profits skyrocketed.

Herbicides, fungicides, and insecticides were soon called "agrochemicals" or more confusingly, "crop protection." Coupled with a seemingly endless supply of the "holy trinity" of nitrogen, potassium, and phosphorus, these new chemical concoctions facilitated never-before-seen quantities of the "commodity" crops of wheat, corn, and eventually soy. Due to the sheer volume of production, the commodity crops were soon turned into animal feed, enabling unprecedented production of beef, pork, and, later, chicken.

The "Corn-CAFO complex" was established. Commodity field

corn in the Midwest was shipped to concentrated animal feeding operations (CAFOs) in places like Kansas, which produced meat for cities on the coasts. More food paved the way for more people.

The United States led the charge in food production and the rest of the world followed. World grain production doubled from 1940 to 1970, then doubled again from 1970 to 2000. But to sustain that growth in monocrop grain production, global pesticide production had to increase much faster than crop yields. Pesticides ratcheted up from almost zero production in 1945 to 500,000 tons in 1955 to over 1 million tons in 1960 to 2 million tons by 1975 to 3 million tons by 1990 to an estimated 5 million tons today.[5]

All seemed well in the world of agriculture. More is, after all, better.

But a look at just a couple of those ten thousand chemical pesticides, herbicides, and fungicides paints a very different picture of what has actually happened with agriculture since the end of World War II.

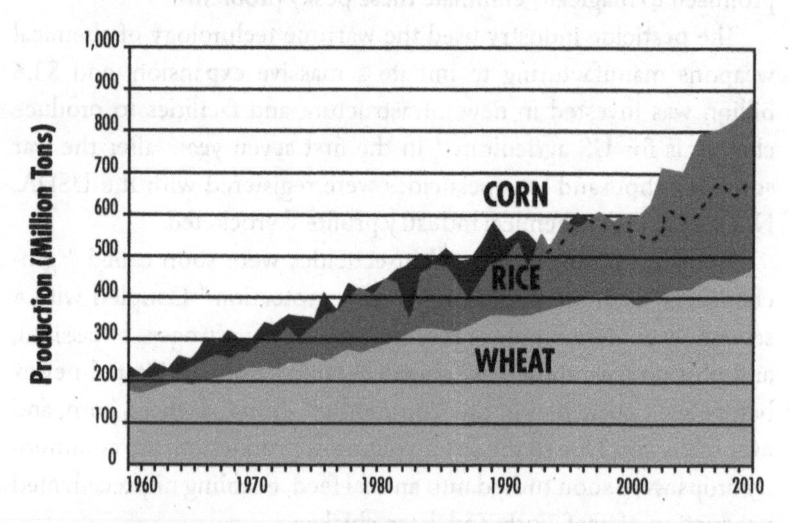

The post-WWII commodity grain production boom led to an overfed and undernourished population.

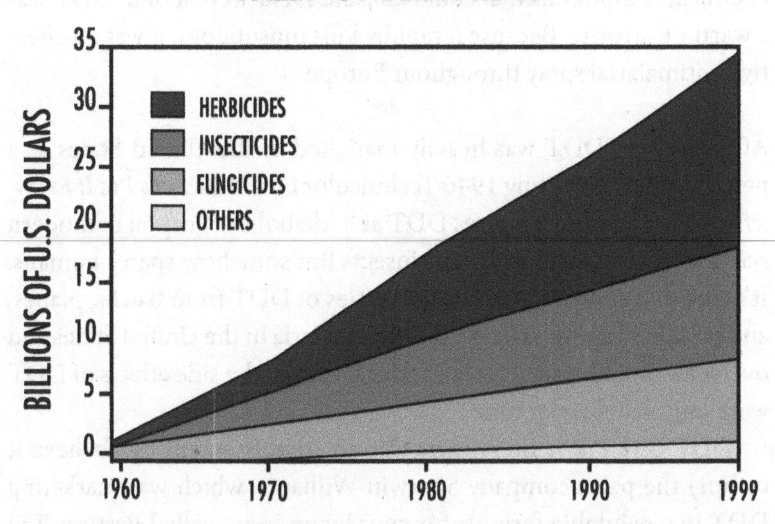

To keep grain production increasing, a far greater increase in toxic chemicals is needed each year. The rate of growth of poisons far outpaces the rate of growth of the grains that require them.
Source: Plant Pathology by George Agrios, 2005.

DDT IS GOOD FOR ME-E-E!

It was a Swiss chemical and pharmaceutical company that developed one of the chemicals that would permeate the American farmscape after the war. Paul Hermann Müller was a chemist working for the company Geigy AG. Müller had lived through a food shortage in Switzerland and wanted to find a chemical that would be better at killing insects than the "expensive natural products" available at the time.[6] He settled on a snazzy compound called *dichlorodiphenyltrichloroethane*, also known as DDT.

Before and during the war, Geigy had factories and subsidiaries in Germany and imported about 80 percent of its raw materials from Germany. While it is unclear if there was a direct connection between the Nazi chemical war machine and the development of DDT, the zeitgeist of the Germanic world in the 1940s centered on developing chemical compounds that could kill, and DDT fit that bill.[7] DDT was a wartime favorite. Because it rapidly kills mosquitoes, it was an effective antimalarial spray throughout Europe.

After the war, DDT was heavily marketed in the United States as a pesticide. In the exciting 1946 Technicolor film *DDT Let's Put It Everywhere,* the announcer praises DDT as a "diabolical weapon of modern science" that miraculously kills insects but somehow spares humans. It's true that spraying massive quantities of DDT from trucks, planes, and handheld sprayers vastly reduced malaria in the United States and overseas. But, like most toxic chemical sprays, the side effects of DDT were only visible over time.

DDT Let's Put It Everywhere was an advertisement by (believe it or not) the paint company Sherwin-Williams, which was marketing DDT in a paintable formula for your living room called Pestroy. The Pestroy campaign was accompanied by a barrage of pro-DDT ads from other companies, including the one that appeared in *Life* magazine for the Pennsylvania Salt Manufacturing Company that depicted a happy chorus of a dog, an apple, a housewife, a cow, a potato, and a chicken all singing "DDT is good for me-e-e!"

After many years of spraying, painting, dumping, ingesting, and coating American homes, streets, and farm fields with DDT, the Environmental Protection Agency (EPA) began to study the chemical in the 1970s. The EPA found that DDT decreases the reproductive rate of birds by causing eggshell thinning and embryo deaths. It is "highly toxic" to fish, which have a "poor ability" to detect DDT in water. Its half-life in water, including the water that runs off farm fields into aquifers, rivers, lakes, and the sea, is about 150 years.[8]

Even though DDT was banned in the United States in 1972, most of us still carry DDT in our blood. It is also still found in rain samples around the world. According to the EPA, which categorizes DDT as a "probable human carcinogen," mammals exposed to DDT develop liver tumors. DDT also tends to accumulate in the fat cells of humans.[9]

Not surprisingly, DDT has been found in high levels of breast milk in the areas where it was heavily sprayed.[10]

In 2015 a study was conducted tracking DDT in fifteen thousand mothers, daughters, and granddaughters across three generations in the San Francisco Bay Area. The study, which was conducted by the Public Health Institute, found that the women with the highest exposure to DDT (which was used heavily in California's farm fields) were 3.7 times more likely to have been diagnosed with breast cancer than women with lower exposure.[11]

DDT is a nasty chemical. But it's a tiny drop in the proverbial bucket of chemicals sprayed onto our farm fields since World War II.

ORANGE IS THE NEW BLACK

At the close of World War II, one of the many chemicals introduced to the American farm was sold under the brand name Weedone. The primary ingredient in this was 2,4-Dichlorophenoxyacetic acid, otherwise known as 2,4-D. Unlike DDT, which was used to kill bugs, 2,4-D is a "selective pesticide" used to kill weeds. It works by being absorbed by the leaves of plants, which then overgrow, wither, and die.

2,4-D is hydrophilic, meaning it attaches to water molecules. This made it the perfect companion in a chemical cocktail developed for another war, the one in Vietnam.

During the conflict, between 1962 and 1971, the US military wanted a way to get rid of all those annoying trees that provided canopy and cover for the enemy. Using the knowledge of "herbicidal weapons" that the United States and Britain had gained during World War II, they combined two chemicals: 2,4-D and something called 2,4,5-Trichlorophenoxyacetic acid, or simply 2,4,5-T. They named the new mixture "Agent Orange."

Theoretically, Agent Orange was to be a "defoliant" to kill the leaves on trees, thus giving military airplanes and helicopters a better view of the villagers below. However, from the inception of the Agent Orange spraying program in 1962, the US military began targeting crops, not trees. During the program, millions gallons of Agent Orange were sprayed over South Vietnam. This led to a widespread famine that left hundreds of thousands of people starving.

The list of possible or probable human health effects of exposure to Agent Orange includes: leukemia, lymphoma, cancer, and heart disease, to name a few. In Vietnam, contested research points to links between Agent Orange exposure and children who are born with cleft palates, mental disabilities, and extra fingers and toes. Numerous lawsuits have been filed against the US government for its use of the toxic compound during the war. While American veterans have won some financial and medical support, the victims in Vietnam have won little more than encouragement to sell more plastic toys to America.

None of this has deterred today's agrochemical companies from selling 2,4-D. Today about seventy-five companies sell herbicides containing 2,4-D under more than one hundred brand names. Their standard response is that it was the 2,4,5-T with its dioxins that was the toxic component of Agent Orange; thus it's fine to continue to spray 2,4-D with abandon over US farm fields, golf courses, parks, playgrounds, and lawns. The World Health Organization (WHO) even published a study saying that "volunteers" who "ingested" 2,4-D passed the majority out as urine after forty-eight hours.[12] (One wonders what method was used to convince participants to "ingest" 2,4-D.) Since then, the WHO now lists 2,4-D as "possibly cancer-causing."

The effects of toxic chemicals in small doses over time is cumulative and therefore must be measured over time. This is why much of today's study of toxicology is based on longitudinal research. When it comes to 2,4-D, the longitudinal research shows a chemical that is far more dangerous than was advertised.

Most of the drinking water tested in the United States has 2,4-D in it. Water supplies have been shut down due to people being able to taste the chemical that 2,4-D turns into (2,4-Dichlorophenol) at as little as 0.3 micrograms per liter. Tested water supplies in the United States, meanwhile, measured between 0.1 and 0.5 micrograms of 2,4-D per liter.[13]

The *Journal of Pesticide Reform* produced a fact sheet in 2005 that details global studies on the effects of 2,4-D. The chemical has a host of unwanted effects, the most disturbing of which are the disruption of the endocrine system, reduced sperm count, damage to testes, and genetic damage. Because 2,4-D is hydrophilic it has a tendency to "stick" to human cells. This chemical also likes to "stick" to other chemicals, including dioxins, which are long-term poisons that do not easily break down in the environment.

2,4-D also transfers directly to breast milk. The chemical is linked with an increased risk of cancer and testicular problems in dogs. And it causes genetic damage to plants even when it is used in such small amounts that it does not visibly damage plants.[14] According to the Centers for Disease Control, about 25 percent of Americans now carry 2,4-D in their bodies.

There's one other thing—it's in our food, too. About 46 million pounds of 2,4-D are sprayed in the United States each year (about 2.3

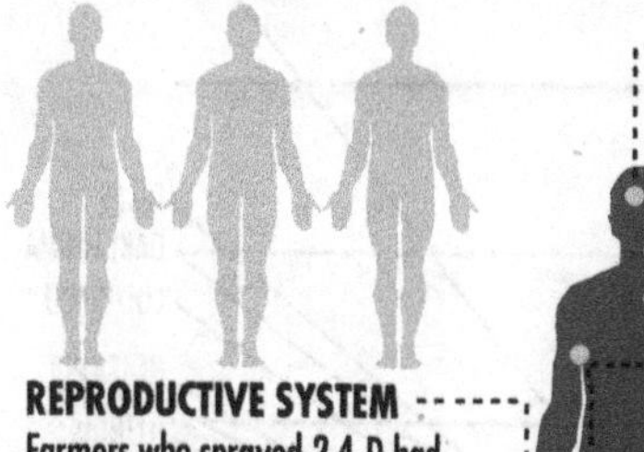

1 IN 4 AMERICANS HAS 2,4-D IN THEIR BODY
PARKINSON'S DISEASE
Human exposure to 2,4-D has been linked to oncreased risk of Parkinson's disease
THYROID
2,4-D has been linked with adverse effects on the thyroid and hypothyroidism in farmers
REPRODUCTIVE SYSTEM
Farmers who sprayed 2,4-D had lower sperm counts, lower sperm motility & greater percentage of anomalous sperm
IMMUNE SYSTEM
2,4-D suppressed immune function in farmers who handled and applied this herbicide
NON-HODGKIN LYMPHOMA
Exposure to 2,4-D & other agricultural pesticides & herbicides has been linked to increased risk of non-Hodgkin lymphoma

Studies link 2,4-D exposure to health problems.

Source: Environmental Working Group (EWG).

2,4-D USE IN AMERICA IS INCREASING

ESTIMATED ANNUAL USE IN 2011

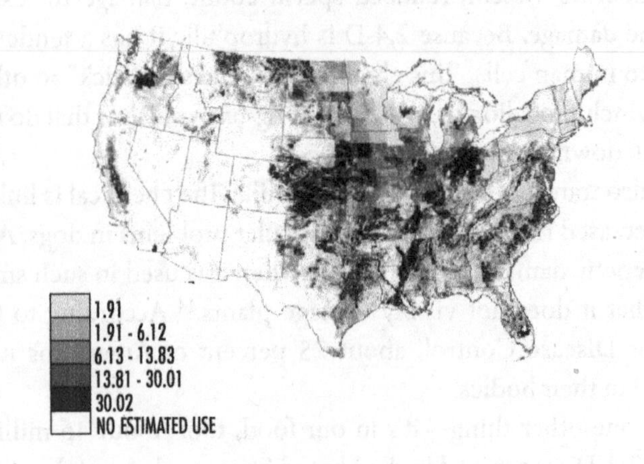

PROJECTED USE BY STATE

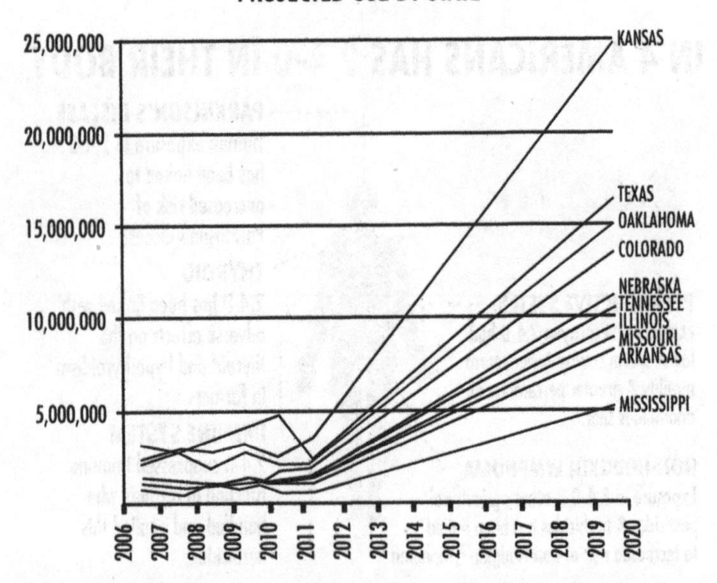

Despite studies that link 2,4-D exposure to a host of health problems, the use of this toxic chemical is projected to rise.
Data: USDA, USGS.

ounces per American per year). The major uses are on pasture and range land (where animals Americans eat graze), on wheat, and on lawns, golf courses, parks, and gardens.[15]

According to Dow AgroSciences the use of 2,4-D is likely to increase because their new Enlist™ brand "weed control system" makes genetically engineered corn and soybeans that are "resistant" to 2,4-D. There's even a slick website dedicated to marketing Enlist. If you're spooked by the thought of more 2,4-D in your food and on your child's playground, don't worry, because the website features a "Certified Advanced Trait Technology" stamp. How reassuring.

According to the site: "Only Enlist Duo® herbicide with Colex-D® technology combines the proven performance of a new 2,4-D and glyphosate. The result: unrivaled weed control designed to land and stay on target. With two modes of action, Enlist Duo eliminates tough yield-robbing broadleaf weeds—including resistant and hard-to-control species. Take control in the field. And take your yield potential to the limit."[16]

Instead of Agent Orange, which combined a dioxin (2,4,5-T) with 2,4-D, this new weed killer combines 2,4-D with something that in the long term may prove just as dangerous.

THE NEW KID ON THE FARM

In the early 1970s Monsanto patented a new weed control molecule (*N*-(phosphonomethyl)glycine) that the company called glyphosate. (The chemical was actually first developed across the border from Germany in Switzerland not long after World War II.)

Glyphosate works by disturbing a plant enzyme critical to the production of amino acids, thereby shutting down plant growth. Glyphosate is also a chelator, meaning it bonds to minerals, binding them and making them unavailable to plants, animals, or humans.

There's a tremendous amount of industry-sponsored science about glyphosate and its primary branded product, Roundup. There's even an industry-sponsored website set up by the Glyphosate Task Force in an attempt to sell the EU on the benefits of this wunder-chemical.

Like most of the other "-cides" developed by the post–World War II chemical industrial complex, the bottom line with glyphosate is predictable. The World Health Organization classifies it as "probably carcinogenic," and the California EPA, not mincing words, just calls it "carcinogenic."

It should come as no surprise that industry-funded studies show glyphosate is not in breast milk while other studies show that it is.[17] The list of probable and possible glyphosate-toxicity issues is the familiar who's who of chemical-induced human health problems, including bioaccumulation, increased breast cancer cells, severe organ damage, mammary tumors, toxic effects to fish livers, and endocrine disruption.[18, 19, 20]

Glyphosate is often sprayed on crops just prior to harvesting. In other words, it's sprayed on foodstuffs just before they are turned into food. It is used primarily on corn and soy, which are genetically altered to withstand multiple sprayings, secondarily on wheat and cotton, and thirdly, on pasture and hay. A USGS map of the 100 million acres onto which glyphosate is sprayed shows that its use corresponds clearly to the "big agricultural" areas of the United States, with the most concentration in the Midwest and down the Mississippi River basin. Another area of concentration runs along the East Coast. Finally, there is a heavy-use patch that corresponds to California's Central Valley growing region.

The use of glyphosate has skyrocketed since the mid-1996 introduction of genetically modified (GMO) crops that are "resistant" to the chemical. All told, around 300 million pounds (about a pound of glyphosate per American per year) is sprayed onto farm fields and city sidewalks across the United States.

This blanketing of America by this bioaccumulating toxin only gets more troubling when one considers that there are now high levels of glyphosate showing up in our organic breakfast foods.[21] Apparently glyphosate is even making its way into our air and water. According to the US Geological Survey, in samples taken along the Mississippi delta

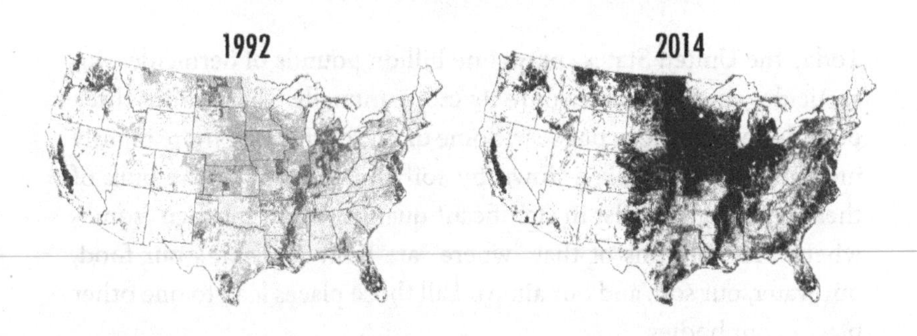

Glyphosate use in America radically increased after the introduction of GMOs and is still expanding throughout the major growing regions.
Source: USGS.

in 1995 and 2007, 86 percent of air samples and 77 percent of the rain samples contained glyphosate.[22]

Let's set aside the possible health effects for a moment. The main problem with glyphosate from the "produce as much as possible" standpoint is that the more glyphosate farmers spray, the less effective it becomes at killing weeds.

Soon after glyphosate began being sprayed in vast quantities new strains of "superweeds" began popping up in exactly the same places the chemical is sprayed. Today there are some sixty-five "confirmed glyphosate-resistant weeds," which include horseweed, common ragweed, giant ragweed, Palmer amaranth, common waterhemp, hairy fleabane, Italian ryegrass, rigid ryegrass, and Johnson grass. About 50 percent of farmers surveyed say they have glyphosate-resistant weeds.[23]

Just when the agrochemical companies thought they could outsmart nature here come all these herbicide-resistant weeds. With debilitating enemies like these what's a chemical company to do?

The answer is simple. Use more.

CHILDREN OF THE CORN

Today the United States sprays one billion pounds of herbicides, insecticides, and fungicides onto its crops annually. That's about three pounds per American per year. Some of these -cides are more rapidly metabolized and broken down by soil and plants. But the bulk of these chemicals persist in significant quantities and must go "somewhere." The options of that "where" are limited. There's our food, our water, our soil, and our air. And all those places lead to one other place—our bodies.

ANNUAL PESTICIDE USE

3 lbs. of pesticides are used per American, per year

Some of the one billion pounds of pesticides, herbicides, and insecticides used in America find their way into our bodies, where they wreak havoc with the endocrine system and can increase the risk of diseases like pediatric cancer.

The second-most-used herbicide after glyphosate is called atrazine. As of late, atrazine has gotten some fairly negative reviews. In 2004 the European Union banned atrazine after finding high levels of it in drinking water. In its study of US groundwater, the US Geological Survey found atrazine is the most common herbicide in groundwater.[24] According to studies published by the EPA and in *The Journal of Steroid Biochemistry and Molecular Biology*, atrazine is an endocrine disruptor.[25, 26] This means it disturbs the hormonal system of the human body. Those most susceptible to endocrine disruption happen to be children.

Like most herbicides, insecticides, and fungicides, atrazine is sold under numerous brand names in bottles and jugs with pictures of healthy-looking flowering plants. Some of these bottled chemicals combine atrazine with other things like organophosphates, dioxins, and yet other chemicals, depending on the targeted weed population for your lawn, park, corn crop, soy crop, et cetera. In fact, most of the -cides sold on hardware store shelves and in farm chemical cooperatives across the United States are "combos." If we are unsure of the long-term health effects of just one chemical poison, it is even more difficult to understand the effects of combining multiple poisons. Unfortunately, the canaries in the coal mine in our national pesticide experiment are likely our kids.

The Center for the Health Assessment of Mothers and Children of Salinas, California (CHAMACOS) is a joint project with UC–Berkeley to study the effects of pesticides and pollutants on pregnant women and young children. They have studied six hundred women's bodies, tracking those same women through pregnancy and birth. Then they studied their children. At age two, the children of the women with the highest pesticides in their blood had the worst mental development and the most cases of developmental disorders. At age five, the children whose mothers had the highest exposure to pesticides had poorer attention spans than the other children. Granted, the women studied live in the agricultural growing region of California's Central Valley. But the CDC has found high levels of pesticides in homes across the country, regardless of proximity to agriculture.[27]

The University of California, Davis performed another study in 2014 that tracked 970 children who were born a mile or less from a farm. The study found that if the mothers of those children lived close to a farm that used organophosphate pesticides, their children were 60 percent more likely to develop autism than those children whose mothers did not live close to treated fields.[28]

Lest we dismiss the effects of poisonous chemicals on children as "a California thing," in June 2015 the Cincinnati Children's Hospital Medical Center, working under the National Institutes of Health, published a study in the journal *Environmental Health*. The study found an association between the use of pyrethroid pesticides and ADHD. Pyrethroids are organic compounds that repel pests. They are the basis of most commonly used household insecticides. They are believed to be harmless to humans and broken apart by sunlight.

The study looked at a cross section of 687 children between the ages of eight and fifteen. Boys with detectable urinary 3-PBA (the breakdown chemical that is the biomarker of exposure to pyrethroids) were three times more likely to have ADHD as compared with those without detectable levels of 3-PBA. Hyperactivity and impulsivity increased by 50 percent for every tenfold increase in 3-PBA in boys.[29] Remember, pyrethroids are common household chemicals, sold under numerous brand names at grocery stores, hardware stores, and gardening stores.

Americans are exposed to one billion pounds of pesticides annually. It appears this is even affecting our children's IQs.

Researchers at Columbia University and the Icahn School of Medicine at Mount Sinai in New York City have been tracking several hundred moms from pregnancy through birth. The mothers were measured for levels of organophosphates. The researchers then evaluated their children's motor skills at one, two, three, and seven years of age. The result? For every increment of increased prenatal exposure to organophosphate chemicals, the children's IQs dropped by 1.4 percent, and their working memory scores dropped by 2.8 percent. The relationship appears to be linear. In other words, the higher the exposure, the greater the damage to the child's abilities.[30]

There are over two hundred recent studies that correlate pesticide exposure with damage to children's lungs, brains, and bodies. These insecticides and herbicides are being linked to childhood leukemia, lymphoma, cancer, and sarcoma. Reduction of cognitive function in children is being linked to exposure to insecticides and herbicides with complex inorganic structures. There are a number of studies that are now drawing a direct line between certain pesticides and babies born with abnormalities.

Pesticides are designed to kill. Every year, especially during the summer planting season, we inject them into our ecosystem at a quantity that no society has ever done. Then these poisons bioaccumulate in plants, fish, animals, and eventually us.

The vast body of evidence shows that over time, these same pesticides are killing us.

The data suggests these poisons are in up to 98 percent of food.[31] Sometimes it's in small doses, sometimes large.[32] The USDA, the agency responsible for testing our food, does not test for the majority of the worst offenders of these poisons (including 2,4-D, glyphosate, or atrazine) in the foods on which they are mostly sprayed (corn, soy, and wheat).[33, 34]

Washington, DC's "revolving door" between big agricultural businesses, the regulatory agencies, and the Senate and House committees that are supposed to oversee them leaves little in the way of citizen protection from these chemicals. With nobody to shield them, Americans are the guinea pigs in the largest chemical experiment humankind has ever undertaken.

We're gambling our lives in a one-billion-pound-per-year-toxic-chemical, 324-million-person roll of the dice and the odds are stacked against us. If you're wondering why such a system continues despite the science, ecology, and economics that show that it is a disaster, look no further than our nation's capital.

A BARREL OF PORK

American taxpayers fund something called federal crop insurance. Issued by the Federal Crop Insurance Corporation (FCIC), which was established in the 1930s to help farmers recover from the Dust Bowl and Great Depression, the FCIC now has a $30 billion line of credit from the US Treasury and offers insurance on roughly one hundred different commodity crops. In principle, the FCIC is there to make sure that American farmers don't go broke. But in practice it has become the single most powerful tool in dictating the continued use of genetically modified seeds (GMOs), the toxic chemical sprays they require, and the soil-destroying practices of modern agriculture.

The idea behind federal crop insurance is that if a farmer's crops are negatively affected by inclement weather, including floods, droughts, or hail, or if the finished crop is worth less than the market value due to a sudden downward shift in price, the government will provide a baseline fallback price, even if the crop is a total failure. This ensures a farmer doesn't go out of business due to circumstances beyond his or her control, and in so doing, supposedly ensures food security for America. Not a bad idea, especially for farmers who only grow one crop on thousands or even tens of thousands of acres.

But the deregulation that plagued America in the 1990s and 2000s and that eventually led to the 2008 financial crash also altered the DNA of crop insurance. Federal crop insurance is now issued through the Risk Management Agency (RMA) of the Department of Agriculture, which sets minimum price insurance for crops (also known as "price guarantees"). Crop insurance is also now issued by private crop insurance brokers. Just like on Wall Street, there are new "insurance products" that can be bought and sold.

Not surprisingly, up to 67 percent of the premiums for crop insurance are paid to private companies directly from the federal government. If all that sounds like mumbo jumbo, the bottom line is that private enterprise is soaking up most of the tax money that is supposed to be paid to farmers, who, due to a overbearing and outdated government crop finance scheme, grow the very crops that make Americans sick.

The system of crop insurance works like this: RMA releases its policy listing crop insurance prices. Based on the list of insured crops a farmer decides what they will grow. A farmer then certifies his or her production by making sure it conforms to the government model. After harvest there's an acreage report. If, as is often the case, the crop produces less than the expected per acre quantity set by the government, the farmer files a loss report. The insurance is calculated and the premium is paid by the federal government (mostly to a private company). The farmer receives his or her "loss payment" and the cycle starts again.

Like so many US government programs, there's a big catch-22. For farmers to receive any form of disaster relief, they must *first* participate in the federal crop insurance program. In other words, if you want the security of government help, the crop insurance program is *mandatory*.

MINIMUM PRICE GUARENTEE
AVERAGE 2016 RMA PRICING

COMMODITY CROP	RATES (PER UNIT)
CORN	$2.90 PER BUSHEL
SOY	$8.90 PER BUSHEL
WHEAT	$5.60 PER BUSHEL
RICE	$11.20 PER 100 LBS
BARLEY	$2.80 PER POUND

2016 Minimum Price Guarantee Average, RMA Pricing.
Source: RMA.

Before a single seed is planted, a farmer must adhere to the federal insurance program's strict guidelines concerning the type of crop to be planted (i.e., the patented seed), the methods used (i.e., chemicals sprayed), as well as when and where the crops are grown. Not surprisingly, farmers generally grow the crops with the highest per acre insurance rates in their area. Because it provides a guaranteed price for crops, the federal crop insurance program tells the majority of America's farmers what to grow and what not to grow.

While it maintains one sort of food security, in its current incarnation the government crop insurance penalizes farmers who do the right thing when it comes to soil. Based in Washington, DC, where the average Senate seat costs around $10 million and where there are over one thousand lobbyists for every member of Congress, the FCIC is in lockstep with the major companies that profit and benefit from industrialized corn and soy and the chemicals and machines they require.

One sometimes hears the federal crop insurance program referred to as a form of "white welfare" but in reality the program does little to subsidize farmers who are going broke in record numbers. Instead, it indirectly subsidizes the corporations that profit from the seeds, fertilizers, and chemicals by ensuring they have a stable and growing market.

THE BIGGEST LOSERS

The fastest-growing and most profitable products in agriculture today aren't food products, but rather herbicides and insecticides. The global use of these -cides increases yearly and is projected to continue to rise exponentially. A recommendation for how much of a given chemical to use per acre may be two to three pounds (for atrazine, for example) or 1.5 pounds (for glyphosate, for example), but as weeds become more resistant farmers who need more punch can, and often do, add more. And despite the promised endless food-production magic of these agrochemicals, the global production output of grains is beginning to stagnate.

From 1950 to 1990, during the "green revolution" of new seeds and chemicals, global grain yields increased by an average of 3.5 percent a year. It literally seemed that the world could keep growing more food per acre forever. But that growth started to level off in the 1990s (just when GMOs and glyphosate were introduced). From 1990 to 2010 annual yields only increased 1.3 percent (a 60 percent deceleration).[35] Meanwhile, the per-capita arable land per person has dropped from around 0.41 hectares per person globally in 1961 to around 0.21 hectares per person today.[36]

In other words, there is less land per person and the "green revolution" that promised never-ending upward growth curves of farm products is reaching the upper limit of how many calories it can pump out of a given acre of land. Things are getting tighter. And no one feels the squeeze more than farmers themselves.

In 2009, the midsize farms—those grossing between $100,000 and $250,000—averaged a net income of approximately $19,270, including government payments. Even those operations designated by the USDA as "large industrial farms" (making a gross income of between $250,000 and $500,000 in 2009) netted only $52,000 on average, including $17,000 in government payments.[37]

A 2015 University of Illinois Department of Agricultural and Consumer Economics (ACE) budget projection puts the net farmer income in 2016 for corn at negative $66 and soybeans at negative $97, respectively. Meaning, growing corn will result in a loss of $66 per acre and soy will lose you $97 per acre. The report's recommendation? Cut costs by $100 per acre. Then at least you could make $3 an acre with soy.[38] In other words, the *only* way for farmers to make any money on these crops is with government insurance money, otherwise known as a subsidy.

With incomes like these, it's no surprise that farmers are leaving the land in droves. According to Farm Aid, 330 American farmers leave their land for good every week. In India the overproduction of farm commodities, overwhelming debt, and pressure to conform to chemical companies' use of expensive pesticides have driven many farmers to commit suicide. Since the early 2000s (when GMOs were first introduced in India), suicide has taken the lives of over one hundred thousand farmers in India. In a cruel twist fit for the likes of Dr. Mengele, one of the most common forms of self-immolation is by drinking pesticide.

But not everyone in agribusiness is struggling. The seven largest pesticide and genetic seed companies are raking in about $93 billion annually. The fertilizer industry is harvesting around $175 billion a year. And all told, the "crop protection" industry of fertilizer, GMO, and chemical pesticide companies is growing nicely with about $350 billion in annual sales, mostly to farmers.[39]

Farmers in the commodity game of "grow more with smaller margins" will find the tightrope they walk is getting more tenuous. Six companies now control 75 percent of the grain-handling facilities, forming a virtual "sextopoly" (which sounds like more fun than it is). As in the days of old, this virtual monopoly of companies sets the price of grain, and farmers have to accept it.

The eating public of America is being ripped off as well. About twenty food companies now produce the vast majority of calories we eat. Their products are primarily bulk foods with low nutritional densities and high profit margins (most of the "stuff" found in a box, bag, or carton). In fact, the majority of Americans get the majority of their calories from Midwest commodity crops (GMO field corn, GMO soy, and wheat), which are sprayed with poisonous chemicals. All of this is courtesy of wartime Nazi technology.

Nutritional density is a difficult thing to access over time, but studies put the loss of nutrition in fruits and vegetables over the past sixty years at anywhere between 5 and 40 percent.[40] Meanwhile, the size of our vegetables, grains, and protein sources has ballooned. This is called the "dilution effect," whereby we eat more calories but receive less in the way of bioavailable nutrients. Perhaps that's why, according to the United Nations, obese people (1.5 billion) now vastly outnumber those who suffer from chronic hunger (925 million).

Because we are visual creatures first and cognitive creatures sec-

AN OBESE
& HUNGRY WORLD

UNDERNOURISHED PEOPLE
925
million

OBESE OR OVERWEIGHT PEOPLE
1.5
billion

FOR EVERY UNDERNOURISHED
PERSON, THERE ARE TWO WHO
ARE OBESE OR OVERWEIGHT

Modern agriculture has led to the advent of the fattest, most hungry global population in human history.

ond, we have traded the substance of food for its look and feel. Unfortunately for humanity, that trade has deadly consequences.

FRACKING FOR FOOD

Once upon a time, nation-states went to extremes in order to obtain the "holy trinity" of nitrogen, phosphorus, and potassium. Today the quantities of these inputs used in the United States alone reach into the trillions of pounds each year. While we are no longer sending ships to far-flung islands to get bird poop, we are now doing something with far greater implications—hydraulic fracturing for nitrogen.

Both phosphorus and potassium are mined in strip mines. While there is some concern about the limits to the supply of these inputs, the immediate environmental ramifications of their mining and use seem a more pressing issue.

The real clincher, though, isn't the phosphorus or the potassium, it's the nitrogen. Because Earth's atmosphere is 80 percent nitrogen, this element is, for the purposes of humanity's needs in the foreseeable future, limitless. Since the Haber–Bosch process just sucks the nitrogen right out of the air, making synthetic nitrogen seems like an open-and-shut case. But making synthetic nitrogen requires hydrogen. Lots of hydrogen. And it also requires lots of energy. That's why the Haber–Bosch process relies on natural gas.

Natural gas (CH4) is a hydrogen-rich gas that is drilled from the Earth's crust. The newest form of gas drilling is called hydraulic fracturing, or fracking, for short. The process involves curving a long drill pipe horizontally under the earth, sending explosives down it, and flushing the well hole with substantial quantities of water. Because demand has been increasing for natural gas for electricity generation and fertilizer production among other things, hydraulic fracturing has risen in popularity in the past decade. Today about 50 percent of the natural gas produced in America comes from frack wells located primarily in North Dakota, Oklahoma, New Mexico, Texas, Louisiana, and California.

Fracking has been linked to tainted drinking water supplies, earthquakes, and extreme environmental degradation. But hydraulic fracturing is the only way America can produce enough natural gas to sustain the factories that make synthetic nitrogen (ammonium nitrate). And without that synthetic nitrogen, more than 90 percent of the crops grown in America would fail.

A 2015 Duke University study found that hydraulic fracturing in the United States used around 250 billion gallons of water from 2009 to 2014, or 27 billion gallons a year.[41] Most of that water is considered "produced" water. Much of it is laced with toxic chemicals and radioactive isotopes. But the petrochemical industry has made numerous assurances that the water is "cleaned," after which they put it back

FRACK WATER PRODUCTION
PRODUCED WATER:
Water seperated from oil during production
Not suitable for public water supply systems
50,000 injection wells in California
RECOVERED OIL AND PRODUCED WATER ARE SEPERATED
PRODUCED WATER IS INJECTED INTO WELL FOR DISPOSAL
PRODUCED WATER IS TREATED FOR AGRICULTURAL USE
EVERY BARREL OF OIL PRODUCED ALSO GENERATES ON AVERAGE
15 BARRELS OF WATER

into use. The good news is this system sounds benign. The bad news? You're probably already eating food grown with frack water.

California farmers have been buying produced water from oil companies for two decades to water crops like citrus and nuts. In Kern County, 21 million gallons of produced water from just one oil refinery feeds forty-five thousand acres of the Central Valley's crops. But now with frack water on the rise, farmers are buying water that has been shown to contain high levels of acetone and methyl chloride as well as oil.[42]

Because California aquifers are being sucked ever lower, produced water from hydraulic fracturing is required to be "reinjected" into groundwater. There are fifty thousand water "disposal wells" in California into which drillers deposit over 20 billion gallons of frack water each year.[43] With lax regulatory standards and even more lax oversight by state and federal testing agencies, the fruit, nuts, and veggies grown in the nation's largest agricultural state (California) are slurping up their fair share of dubious water. If this worries you, think of it this way: polluted groundwater is just a small sacrifice for the greater good of producing synthetic nitrogen, which is critical for crops. Or is it?

More bad news: Most of the synthetic nitrogen applied on farm fields in the United States is not going into crops. Recent studies put nitrogen uptake by crops at about 30 percent. This means 70 percent of what is applied either goes into the atmosphere or into water. Hence, "two-thirds of the US drinking water supply is contaminated at high levels with carcinogenic nitrates or nitrites, almost all from excessive use of synthetic nitrogen fertilizer."[44]

As the city of Des Moines, Iowa, has learned, synthetic nitrogen poisons water supplies. As fishermen in the Gulf of Mexico and other river delta areas are learning, nitrogen-rich agricultural runoff creates an anoxic environment that kills life. The waters affected by poisonous levels of nitrogen are called dead zones because the majority of the food chain is killed off.

To summarize the history of modern chemical agriculture: Nitrogen. War. Pesticides. Machines. Nazis. More strife. Commodity crops. Overproduction. Debt-enslaved farmers. More war. Grain traders.

AGRICULTURAL RUNOFF IN THE U.S.

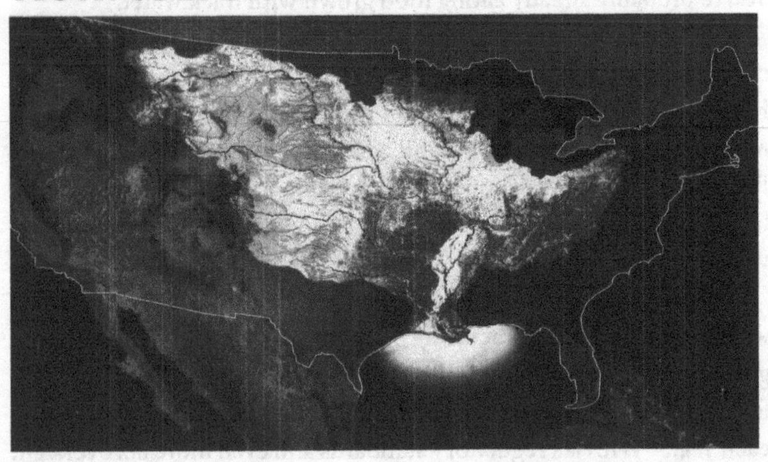

Nitrogen runoff from agriculture across the United States creates anoxic "dead zones" in places like the Gulf of Mexico.
Source: NOAA.

Food monopolies. Sick children. Nitrogen (again). And finally, polluted water. In other words, modern chemical agriculture is a vicious and self-propelled cycle. It answers only to profits and ignores the health of the people for which it purportedly works.

Back on the corn farm I am visiting in Middle America, David the farmer and I are looking out over his fields in the calm twilight. When I ask him about the problems associated with chemical agriculture, he responds with a shrug and says he loves farming. "You'd have to pry this place away from my cold dead hands," he says. "The solitude, the quiet, nature—this is what I love."

This love affair with modern agriculture is, however, a recent tryst. It is the extension of a set of very ancient practices that sealed the fate of nearly every civilization that preceded us. If we are to take a different path forward by reforming how and what we eat, we must go much further back in time—back to when humanity began to farm.

CHAPTER 3

ENDLESS SUMMER

It's an idyllic May day in the southern end of one of the largest food-growing regions in the world. The morning cloud cover is dense, the air is cool, and a light breeze blows.

I'm standing next to a small avocado orchard. It's harvest time. My job is to sort avocados. The big ones that look like candidates for an avocado advertisement go into cavernous forklift-ready bins. The little ones, the discolored ones, and the ones with squirrel bite marks, "the rejects," go into one of several big trash cans.

My wife, Rebecca, and two-year-old daughter, Athena, arrive with a snack. They stay to help. Athena giggles as she tosses the smaller avocadoes I hand to her into the trash cans. Eventually she climbs up inside the big bin of avocadoes that will soon go to a packing house where they will be sorted for size before being shipped off to grocery stores around the country (possibly around the world). It's a picture-perfect moment. A darling little redhead sitting inside an ocean of green, bountiful food. The picture, however, belies a deeper reality.

Today's work on our little five-acre farm is a microcosm of what is being done up and down California so that America, Europe, and countries beyond can enjoy a bounty of fresh produce. Small though it is, ours is counted as one of California's 76,400 farms and ranches that will this year produce over one-third of America's vegetables and two-thirds of the country's fruits and nuts. All told, California produces four hundred critical food "commodities," including the big moneymakers

The author's daughter, Athena, in the Big Picture Ranch avocado orchard in Southern California. (*Simon Balderas*)

of milk, almonds, grapes, beef, strawberries, lettuce, walnuts, tomatoes, and pistachios.[1]

The state also produces most of the country's apricots, dates, figs, kiwifruit, nectarines, olives, and prunes. It is also the nation's leading producer of avocados, lemons, melons, peaches, and plums. We're big on citrus too—only Florida produces more oranges. Then there's the broccoli, carrots, asparagus, cauliflower, celery, garlic, mushrooms, onions, and peppers, all of which the state is a major producer of. There are also animal products. Only Texas produces more livestock.[2] No matter where you are in the world, if you eat any of these foods, you likely eat from our fair state.

Beyond the chemicals that are liberally sprayed on California's farm fields and orchards, and injected into our animals, all this food needs a single critical input that is in ever tightening supply: water.

Like most of the food grown in California, avocados are not native to this area. They come from Central America, a place where it rains every day for three to six months of the year. Thus to grow avocados, we are supposed to use a minimum of 23 gallons of water per day per tree.[3] Our tiny farm has three hundred trees. Watering 365 days a year (times 300 trees, times 23 gallons of water), the recommended amount is 2.5 million gallons of water per year—just for a tiny orchard. Add another 20 percent for irrigation inefficiency, and we're at a staggering 2.75 million gallons of water per year for a five-acre farm.

If our trees were perfect (they're not), they could produce around two hundred avocadoes each per year. On a pound for pound basis (water being 8.3 pounds per gallon), that's 22.8 million pounds of water for a mere sixty thousand pounds of fruit. But even that is not an honest assessment, since a good part of the weight of the avocado is its pit.[4]

What this means in the grand scheme of California's millions of fruit and nut trees and its hundreds of thousands of acres of other crops and animals is that to produce its vast quantities of food, California requires exponentially more water. This is why California uses an almost incomprehensible 38 billion gallons of water per day, about two-thirds of which goes to agriculture.[5] Unfortunately for California and for those of us who like to eat, the water needed to grow our food is going away faster than it is being replenished.

Despite a record-breaking drought the general push is for farmers to keep growing until they strike that perfect alchemy between nature's desires, the cost of inputs (the largest cost of which is water), and the money they make from their produce. Farmers are participating in the equivalent of a casino. Just one more spin of the roulette wheel and maybe this time they'll hit the jackpot.

I ask the lead picker if he's seeing a lot of *pequeños* (small ones) across the Valley. "*Aye, muchos pequeños este ano*" (Yes, lots of small ones this year). He says it's exactly the same situation on every orchard, even the megafarms. Lots more fruit, but much of it cannot be sold due to its diminutive size. I ask him why he thinks this is happening. "*Arboles viejos y poco agua*" (Old trees and little water). A more apt summary might be an outdated agricultural system trying to suck up its last drink.

Since my wife and I can't bear to see the food we grow thrown away, we will redistribute the "rejected" avocados to our community as gifts, we'll save some for ourselves, and we'll have at least one guacamole party. But like the pretty picture of our daughter, a party does nothing to address the underlying issues modern food production faces.

From the inequitable divisions of class, race, and gender of farm labor to the overuse of water to the waste of 30 percent of our food before it even gets to the store,[6] our farm is a microcosm of the larger industrial farming system of California, America, and the world. But the most important thing that's happening on our small farm as well as in agricultural systems around the world is the very thing that eluded almost every civilization practicing agriculture since the invention of the wheel.

We are creating, or rather destroying, our own microclimate.

In other words, the lack of rain, the lack of moisture, the heat coming from the ground, and the stagnant "hot air" that plague agricultural areas worldwide are not accidents. Rather, these seemingly uncontrollable conditions result from how we use our land, the primary use of which is agriculture.

It started a very long time ago.

10,000 BCE

Around twelve thousand years ago in an area known as the Fertile Crescent, early humans began to cultivate grains. This was the beginning of what is called the Neolithic Revolution. The act of "domesticating" plants allowed humans to wean themselves off their hunter-gatherer lifestyles for the more stable existence of towns and eventually cities.

The Fertile Crescent, otherwise known as "the cradle of civilization," is an area that once encompassed what we today know as Iraq, Syria, Lebanon, Jordan, Israel, Egypt, southeastern Turkey, and southwestern Iran. Today the area is largely a dry, arid area characterized by a lack of rainfall and an abundance of sand and rock. Compared to the bounty of the United States, this region offers little in the way of rich soils. But in ancient times the region was lush with life brought by the Tigris and Euphrates and the Nile.

It was here that humans developed early irrigation, the wheel, and writing. This once green and marshy land gave rise to the eight Neolithic founder crops. These crops included the fathering plants to three cereals—emmer wheat, einkorn wheat, and barley, as well as four legumes—lentil, pea, chickpea, and bitter vetch, and one oilseed—flax.[7] It is also the region from which some of the most important species of domesticated animals are descended, namely cows, goats, sheep, and pigs.

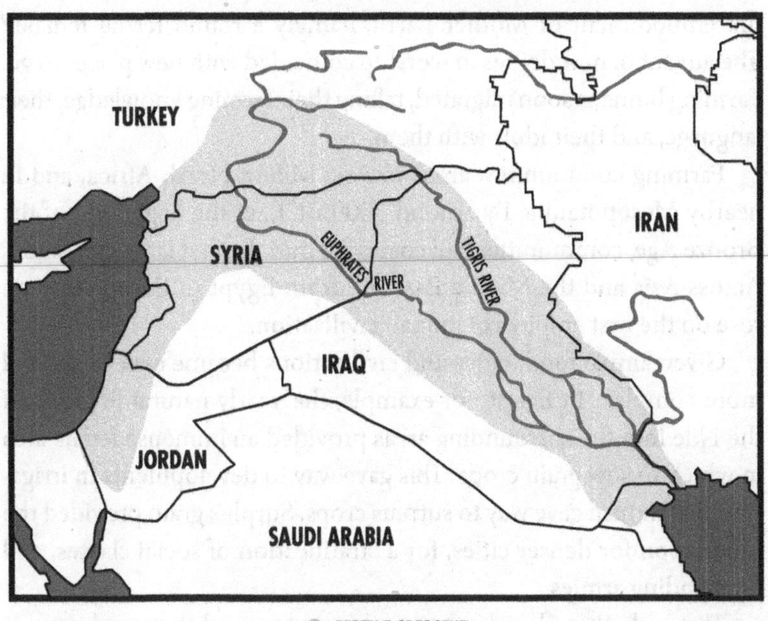

Today the birthplace of agriculture is characterized by desert, lack of rainfall, and destabilized nation-states.

The end of the Pleistocene Epoch (the Ice Age) resulted in the floods, forests, and retreat of bitter cold that created the ecosystems we know today. The retreating ice also coincided with the birth of agriculture. Based on archaeology done at sites like Jericho, the first farmers were likely "camping" in the late spring and summer, during which time they cultivated a limited number of crops. As temperatures warmed, humans were able to stay in the region for longer periods of time and eventually settled into year-round encampments surrounded by farms.[8]

Interestingly around that same time depictions of "goddesses" began to appear in Neolithic culture. These carvings and drawings represent a significant shift for humans, who now began to view themselves as separate from nature. Yet their primary deity was still the embodiment of Mother Earth, namely a round fertile female.[9] The advent of new deities to worship coincided with new places to go. Farming humans soon migrated, taking their farming knowledge, their language, and their idols with them.

Farming communities arose in Asia Minor, North Africa, and in nearby Mesopotamia. By around 3000 BCE, at the beginning of the Bronze Age, communities had coalesced into the first large city-states. Across Asia and the Middle East and from Egypt to Europe the sun rose on the first empires of human civilization.

Given ample food, cities and civilizations became ever larger and more complex. In Egypt, for example, the yearly natural flooding of the Nile into the surrounding areas provided an immense fertile area in which to sow grain crops. This gave way to developments in irrigation that in turn gave way to surplus crops. Surplus grain provided the foundation for denser cities, for a stratification of social classes, and for standing armies.

The cycle that played out over the next several thousand years is writ large across the ruins of the once great metropolises of yesteryear. With a few exceptions, as societies grew, they simultaneously overtaxed their land and needed more food. More food required more land. Neighboring land was the most accessible and also the most likely to be under cultivation by other groups. Standing armies came in handy as empires vied for control of fertile land. Then, just as quickly as they had evolved, most of the city-state civilizations based on large farming enterprises collapsed.

War, famine, flooding, drought, and disease were all factors in the collapse of early civilizations including the Romans and Greeks, and even across the world with the Mesoamerican Inca, Aztec, and Maya. Despite these variables, there was one constant: food. Each society that failed to outsmart nature by producing enough food to keep up with their growing populations faced the same fate. Food is, after all, the basis of civilization. Upsets in the ecosystem, whether human induced or natural, can quickly wipe out an already overtaxed food supply. From there, it was a matter of time before many of the cities of the Bronze Age ended in ruin.

In his infamous twelve-volume opus *A Study of History*, the British historian Arnold J. Toynbee theorized that all civilizations pass through the stages of genesis, growth, time of troubles, a universal state, and disintegration. While not an archaeologist, Toynbee postulated that it was not lack of control over the environment that typically leads to the self-destruction of a civilization, but rather a blinding commitment by the ruling class to outmoded models of problem solving. In other words, "cognitive stagnation."

Whether they fall in love with their own image as leaders or they simply fail to see the writing on the wall, societies get to a point where their leaders squelch or forbid innovative thinking about the problems in their society. From there it is just a hop, skip, and a jump before their empires lay in ruins in the jungle and Indiana Jones is running through the forest with his whip.

Toynbee spent the majority of his life studying the intricate workings of civilizations and he proved his theory time and time again. But added to his theory is the fact that regardless of geography, religion, or politics the central problem for any civilization, the one that requires the most "rethinking" as a society pushes up against its environmental constraints, is how to feed its people. This challenge seems to be the single determining factor in how long a society can last.

DECLINE OF THE EMPIRES

Another man who studied civilizations, not through the lens of political and social structures as had Toynbee, but rather through the lens of land, was the late Rhodes Scholar and soil conservationist Walter Clay Lowdermilk.

On the heels of the debilitating Dust Bowl of the 1930s, the US government sent Lowdermilk and a team of soil and archaeological experts on a multiyear trek around the world. Lowdermilk's mission was to figure out how past civilizations managed their soils and to bring that knowledge back to the United States to help design a "permanent agriculture" system that could last for over one thousand years.

Lowdermilk's land and civilization survey took him through England, Holland, France, Italy, North Africa, Syria, Jordan, Egypt, and the Far East when, after eighteen months, it was interrupted by Germany's invasion of Poland and the start of World War II. Because Lowdermilk and his team of experts were able to study both the land and the archaeological ruins of city-states together as complete systems, they were able to discern a pattern of empire building and collapse that is unmatched in its clarity of cause and effect.

Lowdermilk summarized his findings of ancient empires' land management in a booklet entitled *Conquest of the Land Through Seven Thousand Years*. Millions of copies of the booklet were printed by the US government and it became the inspiration for much of the work of the Department of Agriculture's Soil Conservation Service (which is today called the Natural Resources Conservation Service). In fact, the booklet is still widely available as a free, downloadable PDF.[10]

Of his global study of soil, land, and civilization Lowdermilk said:

> My experience with famines in China taught me that in the last reckoning all things are purchased with food. . . . For even you and I will sell our liberty and more for food, when driven to this tragic choice. There is no substitute for food.
>
> Seeing what we will give up for food, let us look at what food will buy—for money is merely a symbol, a convenience in the exchange of the goods and services that we need and want. Food buys our division of labor that begets our civilization. Not until tillers of soil grew more food than they themselves required were their fellows released to do other tasks than the growing of food—that is, to take part in a division of labor that became more complex with the advance of civilization. . . .
>
> Food comes from the holy earth. The land with its waters gives us nourishment. The earth rewards richly the knowing and

> diligent but punishes inexorably the ignorant and slothful. This partnership of land and farmer is the rock foundation of our complex social structure.[11]

Lowdermilk went on to explain that as civilizations rose, their first order of business was often to cut down trees. Trees and their wood provided for the physical edifices of a society: the ships, buildings, firewood, ash, chariots, and carts. From there land was cleared, agriculture was practiced, and water was channeled to create irrigation. By opening the land and ridding it of the protection of trees and vegetation, the soil was exposed to erosion by wind and water.

Lowdermilk cataloged hundreds of civilizations devastated by the same fate. Time and time again, erosion of the precious soil that sustained agriculture left cities, states, and empires struggling to farm. As low-lying farmland was turned into rock and gorges, trees were often cleared on mountain slopes, exposing the soil on mountains to ever more erosion. The resulting silt from rain runoff would then disturb waterways, blocking the flow of precious water to crops. Unbeknownst to their leaders and their farmers, these civilizations were dabbling in humanity's first wave of experiments with desertification.

If the Bronze Age was an experiment in desertification version 1.0, the Iron Age would see the second wave. But these civilizations were localized and their resource bases were relatively unshared. We are now somewhere around version 3.0 of the desertification experiment. And this time, our civilization is global, our resources are shared, and the stakes are enormous.

DESERTIFICATION

Desertification is a term that applies to the man-made destruction of the underlying soil resource to such a degree that it can no longer support agriculture.[12] In very general terms this human-induced phenomenon results from overcropping, overgrazing, and overusing land due to increased population pressure.[13]

A closer look at the mechanisms that cause fertile land to turn to desert shows there are three distinct phases to desertification:

Phase 1: Disturbing the local water cycle
Phase 2: Increasing the heat of the ground and the air
Phase 3: Topsoil erosion

Phase 1 begins when land is cleared for urbanization, suburban sprawl, and/or agriculture. In this phase, natural ecosystems are cleared to build roads, buildings, freeways, open farm fields, and rangeland. Meanwhile, drainage systems are installed to move water off the land.

Prior to this phase, natural biological systems of trees, shrubs, grasses, and other types of flora serve a number of important functions, including cycling water and protecting soil from solar radiation. The sun, it turns out, doesn't actually send heat to Earth (as we generally think). Instead, it emits radiation. Given Earth's ecosystems, that radiation has three potential pathways when it hits our planet: it can be transformed into photosynthesis by plants; it can be transformed into evaporation; or it can be transformed into heat.

Dense plant cover such as a multispecies ecosystem in a forest, shrubland, or even a savanna or tallgrass prairie will absorb about 80 percent of the sun's radiation and use that energy for photosynthesis and evapotranspiration (sweating out water). The remaining solar radiation not absorbed by plants will be reflected back into the atmosphere and transformed into heat.[14]

In a square meter of forested or heavily planted area, around three liters of water will be evapotranspired into the air each day. This creates a "swamp cooler" effect whereby plants drop the localized air temperature through evaporative cooling.[15] But not all the water that evaporates from the leaves of plants stays at or near ground level. Some of it goes into the atmosphere as well.

In a functioning ecosystem, about half the precipitation that falls on land is derived from water that evaporated from the land (as opposed to the oceans). This land-to-land precipitation-evaporation-

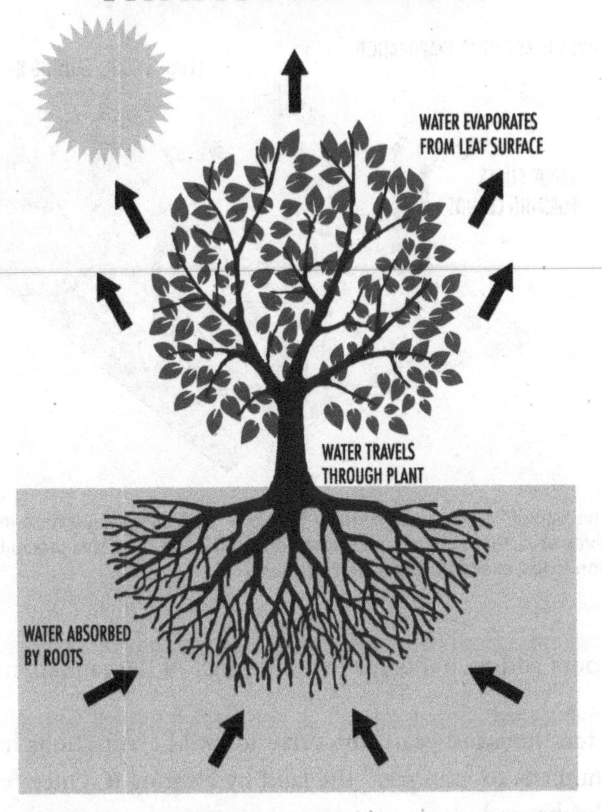
TRANSPIRATION
WATER EVAPORATES FROM LEAF SURFACE
WATER TRAVELS THROUGH PLANT
WATER ABSORBED BY ROOTS

Transpiration through plants cools the air and contributes to mists, dews, and fogs. Collectively, the transpiration from plants creates the "small" water cycle.

precipitation cycle is referred to as the "small" or "short" water cycle. The precipitation of the small water cycle is characterized by light, frequent rainfall, mist, dew, and short showers. It's the kind of precipitation that builds vegetation and helps us grow food.

In contrast, the "large" water cycle is the one we typically learn about in grade school, wherein water evaporates from oceans, forms clouds, and falls on land. This water cycle is characterized by heavy

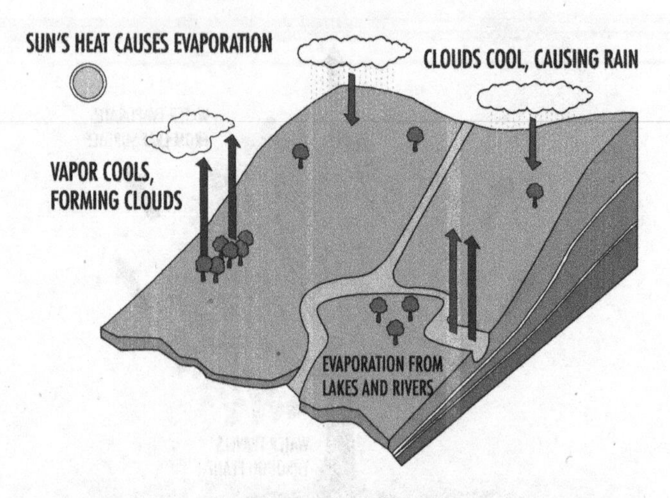
SMALL WATER CYCLE
SUN'S HEAT CAUSES EVAPORATION
CLOUDS COOL, CAUSING RAIN
VAPOR COOLS,
FORMING CLOUDS
EVAPORATION FROM
LAKES AND RIVERS

The "small" water cycle contributes up to half the precipitation in a given area. When plants are cleared and soil is left bare, this precipitation stops, exacerbating desertification.

downpours and accounts for the other half of the rainfall in a given area.

For ten thousand years, the drive to build civilizations has compelled humans to "conquer" the land by clearing it. Once removed, trees, shrubs, and plants cannot evapotranspire water, and the short water cycle begins to break down. Added to this is our need to "drain" land with canals, pipes, and man-altered rivers. "Getting rid" of the water in a given area leads to less absorption by the soil, which means less water is available to evaporate. Once this process begins, it becomes a feedback loop. Each successive rainfall is less "effective" than the last.

Phase 2 of desertification begins as soon as a city or a farming area has been cleared of most of its vegetation and all that remains are its hard, impermeable surfaces (cement, asphalt, roofs, or bare, hardened ground). Since the solar radiation that hits these areas cannot be transformed to evaporation and it cannot be transformed into photosynthesis, it has only one pathway: heat. As cities and mostly barren agricultural fields expand, they transform the tremendous energy of the sun into heat. The larger the impermeable area, the hotter the region gets.

These "heat islands" cause a divergence of air temperatures between day and night and also between one region and the next. As hot air currents above a heat island increase, water vapor is quickly whisked away toward regions with vegetation or higher altitude.[16] Over time, the light, soft, frequent rainfall of the short water cycle becomes extremely infrequent. Heat islands are instead characterized by seasonal and cyclical torrential downpours from the large water cycle. Thus heat islands are also prone to flooding and soil erosion.

According to research done by the EPA and the California EPA, cities with 1 million people or more become "urban heat islands" in which every additional 1 million inhabitants basically equates to a 1 degree Fahrenheit rise in temperature over surrounding regions.[17, 18] All that

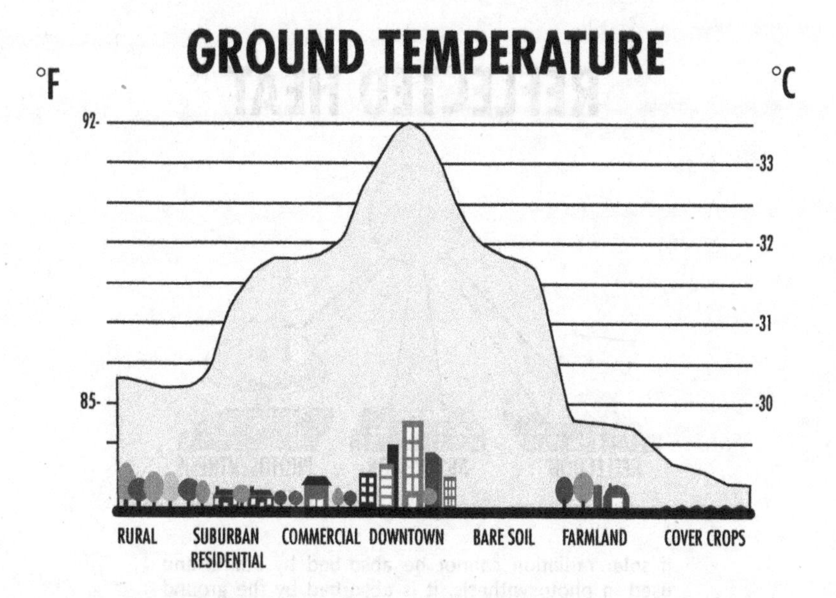

The surface temperature of any hard surface, including bare soil, is much higher than the surface temperature of plants.

hot air turns into heat vortices that essentially "push" the rain away. NASA has measured this effect with its Tropical Rainfall Measuring Mission (TRMM) satellite, showing rainfall rates are up to 50 percent greater downwind of cities.

A dry, hard surface reflects up to 35 percent of solar radiation back into the atmosphere. This "reflected solar energy" heats the air. The dry surface will then absorb the remaining solar energy (about 60 percent of the original solar radiation), transforming it into heat that initiates a baking action. (Think hot plate.) This "hot surface" heat is the type of heat you can feel. Thus it is called sensible heat. So, in Phase 2, the ground and the air above it become increasingly hotter. This results in less rainfall and less frequent rainfall.

Phase 3 of desertification begins once soil is laid bare, but its effects typically take more time to appear. Without plants to keep it cool and with increasing surrounding temperatures, soil dries out and loses its moisture. What was once rich topsoil teeming with

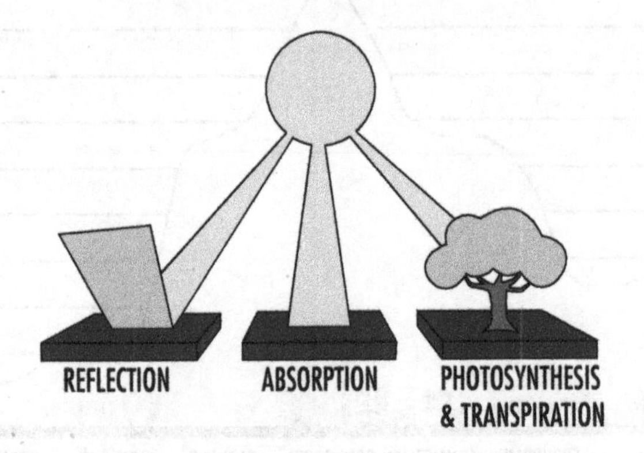

If solar radiation cannot be absorbed by plants and used in photosynthesis, it is absorbed by the ground or reflected back into the atmosphere, creating heat vortices.

microbial life can be reduced to dust in a matter of years, sometimes months. Hotter man-made microclimates result in heavy deluge-type rainfall events and stronger winds, all of which carry away topsoil. This erosion of the basic commodity needed to sustain life is the final phase of desertification. It is also the inevitable result of the way man has practiced agriculture for ten millennia.

Contrary to popular belief, desertification does not only occur in or around a desert; instead, it can occur anywhere. Nor is desertification a result of drought, although droughts can exacerbate the process. Instead, desertification is a man-made phenomenon. This excludes the great deserts left after the ice sheets began to retreat about fifteen thousand years ago. We are focusing instead on once productive areas of the planet that, through the intervention of human civilization, have turned to deserts. This is why the civilizations Lowdermilk and his team surveyed that did not practice soil conservation ended in dust.

The way humans generally practice agriculture pushes land through a progressively worsening desertification. According to Texas Tech botanist Dr. Harold Drenge, the four levels of desertification are:

Slight: Little to no degradation to the soil and plant cover
Moderate: Less than 50 percent of the original plant species remain, or 25–75 percent of the topsoil is lost, or soil salinity (salt levels) has reduced crop yields by 10 to 50 percent
Severe: Less than 25 percent of the original plant species remain, or erosion has removed practically all of the topsoil, or salinity has reduced crop yields by more than 50 percent
Very Severe: Less than 10 percent of the original plant species remain, or the land has sand dunes or deep gullies, or salt crusts have developed on irrigated soils

Today more than 23 percent of Earth's landmass has been degraded by desertification and 1.5 billion people are affected.[19] Africa, Australia, and South America all hover around 20 to 25 percent desertification. About half of Asia is moderately desertified. And, daunt-

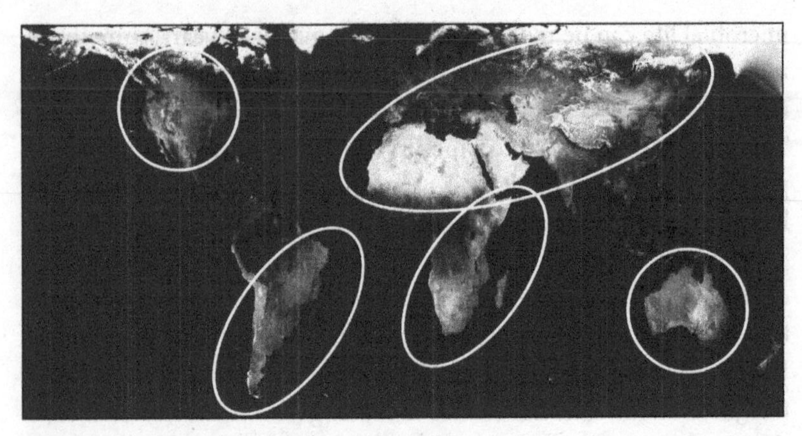

Over two-thirds of the land on Earth is desertifying.
Source: NASA, Savory Institute.

ingly, up to 90 percent of North American arid lands are moderately and severely desertified.[20]

Among the parts of the United States affected by severe desertification are the midwestern growing region, the eastern growing region, the southern growing region, and California's Central Valley. In other words, the places where we grow food. To see desertification, drive to an agricultural area, get out of your car, walk into a field, and look down.

There, between the brush or crops you will see it—dry, bare, eroding ground.

THE BIG SQUEEZE

On a chilly February day in 2014 President Obama boarded the modified Boeing 747 known as Air Force One and flew to California. It was a busy time of year. His trip to the Central Valley was squeezed in between a talk he gave about immigration and the minimum wage at a conference in Maryland and his meeting with King Abdullah II of Jordan in Palm Springs to discuss the Israeli-Palestinian situation and the Syrian Civil War. But the bulk of his three days on the ground in our sunny state was set aside to meet with farmers and stakeholders about the drought.

On the first day of his trip, President Obama teamed up with Governor Jerry Brown and Department of Agriculture secretary Tom Vilsack. Together, they traveled to a model California farm and held a press conference. Obama donned his standard crisp white button-down shirt, took center stage standing at a podium in the middle of a bone-dry field, and announced $100 million in loans and other farmer funding for emergency drought relief.

According to President Obama, "Right now, almost ninety-nine percent of California is drier than normal—and the winter snowpack that provides much of your water far into the summer is much smaller than normal." While this was disconcerting for those of us living in California, it certainly came as no surprise.

But then he brought his speech to the average man on the street by saying, "California is our biggest economy, California is our biggest agricultural producer, so what happens here matters to every working American, right down to the cost of food that you put on your table." This was the hot button he'd been carefully waiting to push. California's drought could affect the price of food on *your* table. Like all good politicians, Obama knows that low-cost, highly available food is the keystone of a stable society. So if California needed money for water so Americans could get cheap food, then by all means, open the government money hose.

Midway through his speech, Obama masterfully switched gears and began speaking about climate change. He said, "Scientists will debate whether a particular storm or drought reflects patterns of climate change. But one thing that is undeniable is that changing temperatures influence drought in at least three ways: Number one, more rain falls in extreme downpours—so more water is lost to runoff than captured for use. Number two, more precipitation in the mountains falls as rain rather than snow—so rivers run dry earlier in the year. Number three, soil and reservoirs lose more water to evaporation year-round."

President Obama's speech on the California drought is an example of our society's understanding of *what* is happening and our confusion as to *why* it is happening. As in civilizations past, the effects are clear but the man-made triggers elude us. The president accurately described desertification and its disruption of the small water cycle, but he incorrectly pinned their cause on carbon dioxide emissions. It is indeed true that higher temperatures resulting from more CO_2 in the atmosphere can exacerbate man-made water cycle disruption and desertification. It is also true that higher temperatures often lead to less snow and therefore to more rainfall. But CO_2 emissions are not the root cause of California's worsening drought.

Think of it this way, in the middle of California is the Central Valley, a 60,000-square-mile, mostly flat piece of open arid land that has been cleared of its former ecosystem in order to perform large-scale agriculture. A small fraction of it is covered in plants and water (if we're being generous, say 10 percent). The remainder (say 90 percent) is bare, hardened, mostly unplanted dirt and reflective urban/suburban surfaces. During the day it reflects heat into the atmosphere and absorbs heat, baking like a massive frying pan. All the while it is pushing clouds and rain away. Meanwhile its soils are eroding, which means food growers require ever more water. Now add to this man-made desertification the occurrence of season after season of CO_2-assisted heat waves.

In contrast, the former ecosystem of the Central Valley, a mixed oak forest with redwoods to the north and bushland savanna to the south, held most of its water in its soils. In natural forest ecosystems soil organic matter (SOM, aka "organic matter") is anywhere from 3 to 7 percent. For each 1 percent of organic matter, an acre of ground will hold around 25,000 gallons of water in its soils.

So before man practiced agriculture in California, the soil was likely holding around 100,000 gallons of water per acre. Growing on top of that soil was a mixture of vegetation types, all of which pulled CO_2 from the atmosphere and pumped it into the ground where microbes used that carbon to build the very pore spaces that stored the water inside the soil. This is the ecosystem we removed.

Today in California as in other agricultural regions, climate change does indeed play a role in tightening water supplies, but the type of

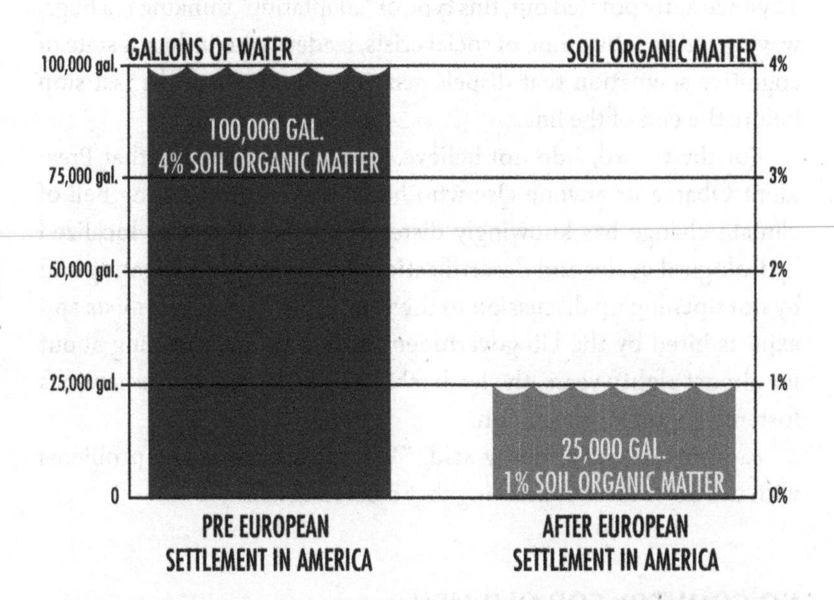

The amount of water soil stores is a function of the organic matter (which is 50 percent carbon by weight) contained within it. The more organic matter, the more carbon in the soil, and thus the more water the soil can store.

change that took the greatest toll was the initial alteration to the land itself.

Herein lies the difference between the micro and the macro or rather between the local climate and the global climate. It is also the difference between something that we feel we can change versus a system that is already set in motion. At the micro level, we deal with regional climate, even the climate of a farm or a watershed (a "microclimate"). At the macro level, we are left to struggle with the atmosphere of the entire planet, something no farmer or city dweller feels much power over.

This powerlessness is why the rest of Obama's speech that February day was about "adaptation," which simply means trying to survive and practice agriculture inside ever worsening conditions. This is what the $100 million in emergency funding was for—adaptation. As Toynbee aptly pointed out, this type of "adaptation" thinking is a huge, waving red flag. In a time of social crisis, leaders who induce a state of cognitive stagnation that dispels genuine solutions are the last stop before the end of the line.

For the record, I do not believe, nor am I suggesting, that President Obama or anyone else who has sounded the warning bell of climate change has knowingly disregarded the impact of localized hydrological cycles and desertification. But by not investigating and by not opening up discussion to the very thing that agronomists and experts hired by the US government itself have been talking about for almost eighty years, the leadership of the climate movement has fostered cognitive stagnation.

As Einstein purportedly said, "We cannot solve our problems with the same level of thinking that created them."

NO COUNTRY FOR OLD MEN

As I drive California's Interstate 5, I see familiar sights: fast-food billboards, dusty fields, and large farmer-made signs and banners. These particular signs have been a mainstay on this strip of California freeway for years. They are slung on sides of semitruck containers, affixed to posts, and strapped to old pieces of farm equipment. They are printed with monikers like FOOD GROWS WHERE WATER FLOWS, NO WATER = LOST JOBS, STOP THE COSTA-PELOSI-BOXER DUST BOWL (referring to California's past and present representatives and senators), and NO WATER = HIGHER FOOD COST! Often perceived as a cry for help, these signs may be better interpreted as the death rattle of a food system that is long past its prime.

My journey takes me eastward off the security of I-5's asphalt and toward the center of California. Hugged by the awe-inspiring bulk of the Sierra Nevada mountains toward the east, the Central Valley watershed comprises over a third of California (60,000 square miles). It is so large that it stretches across eighteen counties and encompasses thirteen metropolitan areas in which some 6.5 million people dwell. But before the Europeans arrived, the valley encompassed one of the largest wetlands in the world.

From its northernmost border to southern frontier with Mexico, the Central Valley was subtropical grasslands, savannas, shrubland, marsh, woodlands, and lakes. Accounts of early settlers tell of estuaries that stretched to the horizon so thick with birds that it looked as if you could walk across their backs. Two million acres of redwood trees buffeted the coast and northern exposure of the valley. Spanish settlers told of having to push through herds of various four-legged animals because they were so tame.

Surrounding live oaks were fountains of lilies that were cultivated by the Miwok and other Native Americans for food. There are tales of salmon running in the rivers so thick you could catch them with a net. Indeed, what is still considered today the most biologically diverse state in the Union was just two hundred years ago one of the most vital ecosystems on Earth.

As I drive, what stretches out before me has clearly been transformed by man. While not as uniform as the Midwest, the desolate farm fields of the valley are largely void of trees and shrubs. Other than what is being planted and grown, this place is also devoid of animal life. An occasional egret or crow can be seen sitting among the lettuce, broccoli, or wheat fields, but there is little in the way of free-ranging four-legged creatures.

Dotted across the landscape like chickenpox are alfalfa and grain-fed dairies. Alfalfa is the largest water-use crop in California and milk is one of California's big agricultural moneymaking exports. Thus, these dairies represent yet another way of converting precious water to cash. The stench comes on strong and increases until the cowpens are visible. It is still early in the summer but as the heat in this flatland increases, so too will the pungent smell. These pens are an affront to the natural majesty of the herds that once roamed this land.

Every twenty miles or so, I cross a cement irrigation canal. These canals are part of California's aqueduct system that provides the life-blood to megafarms. At over 700 miles in total length, California's is the second-longest network of aqueducts in the world. (The longest network of aqueducts, known as levadas, is on the Portuguese island of Madeira and runs 1,350 miles.) The California Aqueduct, with its forty-foot-wide cement troughs, distributes the water from the heavy rainfall area up north around Shasta Lake to where the thirsty crops are in the Central Valley. Because that heavy northern rainfall occurs mostly in the winter and irrigation is needed mostly in the summer, these man-made waterways and dams allow rainwater to be "time-shifted."

In other words, the California Aqueduct is an IV for the nation's biggest agricultural economy.

The use of aqueducts is as old as agriculture itself. They were central to the Bronze Age agriculture of civilizations in Babylon, Assyria, Egypt, and later in Rome. But these waterways were always weak links. Often they filled with silt from upstream soil erosion and stopped flowing altogether. Without water, crops withered and so did the civilizations dependent on them.

But nowhere in the world and at no time in history has one aqueduct supported the life of so many millions of humans as the California Aqueduct.

The quantity of water needed to grow our food is staggering. It takes 5.4 gallons of water for one head of broccoli, 4.9 gallons of water for a walnut, 3.5 gallons for a head of lettuce, 3.3 gallons for one tomato, 1.1 gallons for an almond, 0.75 gallons for a pistachio, 0.4 gallons for a strawberry, and 0.3 gallons for a grape.[21]

If you think that's bad, how about your clothes? California used to be one of the big growers of cotton in the United States. It takes about 1,800 gallons of water to produce the cotton for one pair of blue jeans.

Despite the critical importance of this calorie-producing region, Google Maps has only a frail grasp of the roads that crisscross the vast acreage of the Central Valley. The closest town to where I need to go is Firebaugh, but that only serves as a dusty guidepost indicating the edge of civilization. Beyond it are fields upon fields of workers bent over the ground picking, pruning, and moving waterlines. Their brown faces zip by me, blurring with the brown dirt as the car pummels down the asphalt strip that divides one type of crop from another.

There is nothing romantic about this place. Save the mountains in the distance, there is nothing picturesque here either. It smells of fertilizer and chemicals. It looks like a project on the surface of the moon—barren save for the crops propped up by irrigation systems, which stretch far toward the edge of the horizon. As I drive, I try to push down an impending sense of guilt. Whether I eat quinoa, tacos, or hamburgers, this is what it looks like to grow my food.

THE LAND IS FALLING

By the time I find what looks like a white tin outhouse in the middle of a dirt field, the sun is baking everything in sight. Parked next to the white structure is a white 1990s vintage Jeep Grand Cherokee with US government plates. Hearing my vehicle approach on the cracked earth, a woman in a green T-shirt pokes her head out of the door of the little building. This is Michelle Sneed, a researcher for the US Geological Survey, aka "the USGS."

As US government agencies go, the USGS is a pretty cool organization. It was formed at the behest of the National Academy of Sciences in 1879 as a last-minute addition to an act of Congress to basically study land. According to their website, today the USGS provides "science about the natural hazards that threaten lives and livelihoods, the water, energy, minerals, and other natural resources we rely on, the health of ecosystems and the environment, and the impacts of climate and land-use change." Like I said, it's a cool agency.

Many of the people who work for the USGS, like Michelle Sneed, might be considered "science geeks." Sneed has a firm handshake, is in her forties, and has no kids. "I have a husband and two dogs," she explains. "So I have three dogs," she adds with a smirk. She explains that as an employee of the USGS she is not allowed to give opinions, only report data. At least she has a sense of humor.

Sneed specializes in something called "land subsidence," which means exactly what it sounds like—falling land. It turns out that as early as the 1930s the land surveyors who were staking out the critical waterways that would soon allow California's agricultural boom noticed something odd. Their surveys weren't "closing." In other words, the survey marks they gathered in the field and wrote down seemed to "move" when they returned. Something odd was going on with California's land.

Since that time, land in the Central Valley has fallen and fallen and fallen. In most locations, it's only fallen a foot over the past few decades. But in other locations, it has fallen as much as twenty-nine feet, about a foot a year. The spot where Michelle and I are standing, which is right next to the critical irrigation waterway known as the Delta-Mendota Canal, has fallen twelve feet since the original surveys were done in the thirties.

While land subsidence is a strange phenomenon, it's what is going on under the surface that matters. Aquifers, explains Sneed, are like giant sponges. As we draw water out of them their clay particles squeeze together. The problem is that once they squeeze too closely together, they will never go back to their spongelike structure. Thus an aquifer that has been severely drained of its water can never be fully "recharged." Land subsidence is therefore a way to gauge the severity of underground water depletion over time.

In the case of California, the drought has prompted state officials in Sacramento to severely ration the water coming through the canals that are fed from Shasta Lake in the north. With increasing demand for food and less water, farmers in the Central Valley have had to drill deeper and deeper to get their water from wells. They are then locked into a feedback loop whereby the more water they suck out of aquifers for crops, the lower the aquifers get, making the need for water greater and making farmers drill ever deeper wells. The drought exacerbates the ongoing process of desertification, which further dries the soil. Added to that is the land subsidence, which also contributes to an increase in the soil's salinity, thereby making the yearly need for water even greater.

Land subsidence isn't just happening in California. It's a global

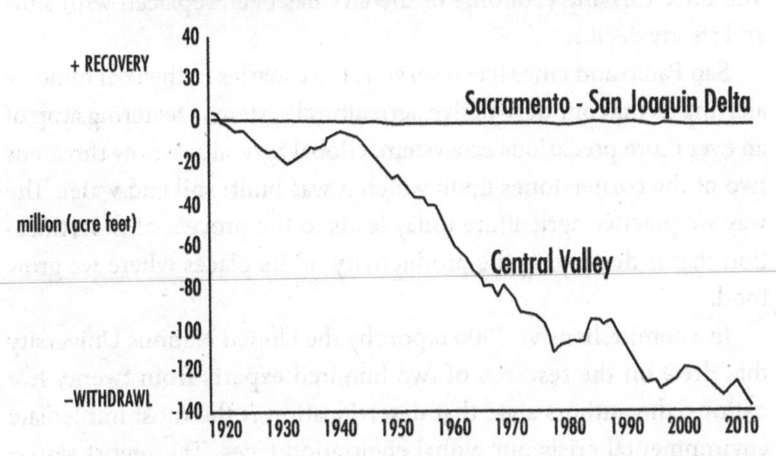

Despite California's long-awaited 2017 downpours, the water level in the state's precious aquifers continues to drop.
Data: USGS.

phenomenon that is occurring in most of the agriculturally productive regions of the planet. Recent NASA satellite images show that twenty-one of the world's thirty-seven largest aquifers have passed the tipping point, meaning more water was removed during the last decade than was replaced. Thirteen of those aquifers are declining at rates that put them into a critical category. America's most distressed underground water supply, not surprisingly, is California's Central Valley aquifer.[22]

In very general terms we are running out of the very thing needed to perform the agriculture that sustains humanity—underground fresh water.

REVERSE TERRAFORMING

In the twelve-million-person city of São Paulo, Brazil, water supplies have become so constricted that tap water is now often rationed. Many who could afford to leave have done so, taking their assets with them. The once thriving economy of the city has been replaced with slow and steady decline.

São Paulo and cities like it serve as the canaries in the coal mine by alerting us that our destructive agricultural system is teetering atop of an ever more precarious ecosystem. Global agriculture now threatens two of the cornerstones upon which it was built: soil and water. The way we practice agriculture today leads to the process of desertification that is destroying the productivity of the places where we grow food.

In a comprehensive 2006 report by the United Nations University that drew on the research of two hundred experts from twenty-five nations, the authors state that desertification is the most immediate environmental crisis our global civilization faces. The report states: "Recent evaluations, such as the Millennium Ecosystem Assessment, clearly demonstrate that there are no signs that the desertification trends are abating on a global scale."[23]

The report goes on to warn that, left unchecked, desertification will likely force millions of people from their homes. Unfortunately, in the ten years since the report was published, it has only become more prescient. Nowhere is this clearer than across the area that was the cradle of civilization.

In his December 2015 piece "The Ominous Story of Syria's Climate Refugees," *Scientific American* writer John Wendle paints a grim picture of the climatic events leading to Syria's current crisis. Writes Wendle:

> Climatologists say Syria is a grim preview of what could be in store for the larger Middle East, the Mediterranean and other parts of the world. The drought, they maintain, was exacerbated by climate change. The Fertile Crescent—the birthplace of agriculture some 12,000 years ago—is drying out. Syria's drought has destroyed crops, killed livestock and displaced as many as 1.5 million Syrian farmers. In the process, it touched off the social turmoil that burst into civil war.[24]

According to the UN Refugee Agency, in 2014, the agency assisted about 47 million people globally, a record high.[25] But even the agency admits that its reach isn't enough to cope with the almost 60 million people who were forcibly moved from their homes that same year. Each year, 1 in every 122 people becomes a refugee and that number is increasing. According to António Guterres, former UN high commissioner for refugees, "We no longer have the capacity to pick up the pieces."[26]

It is no accident that the most severe displacement is occurring across the region where agriculture was first practiced. While there are many factors in the refugee crisis, one cannot unlink the lack of food and water in the region with social disintegration. The area encompassed by the Fertile Crescent has become a worst-case scenario for desertification.

In a twist of fate, we humans "reverse terraformed" our planet from a web of balanced, interlocking, lush ecosystems into larger and vaster deserts. While Syria's climate refugee crisis may be grabbing headlines today, California's Central Valley food-growing region is a potential contender for the headlines of tomorrow. And we are not alone. The nexus of agriculture, water shortage, and human suffering is writ large across India, China, Africa, and even in wet climates like Brazil.

The ancient city-states of the Bronze Age may be dead but they still speak to us. Walk among their rubble and you'll hear the clear warning bells from the many civilizations that successfully practiced (for a time) desertification-based agriculture. If they could speak, the crumbling walls of their palaces and temples might tell us to beware of our hubris.

We are now disrupting the small water cycle, desertifying the land, and causing soil erosion on the scale of a megalopolis. Think Los Angeles, New York City, Shanghai, Delhi, et cetera. With each megacity comes exponentially bigger megafarming. Today the bare, open land spaces around a city stretch for thousands or tens of thousands of square miles in every available direction. Think Central Valley.

Our modern agricultural system may produce more calories than ever before, but our way of growing food is a limited-time offer. Just as in civilizations past, the very food system that gave rise to our modern society now tilts on an ever-eroding foundation. Lowdermilk put it this way:

> Here in a nutshell, so to speak, we have the underlying hazard of civilization. By clearing and cultivating sloping lands—for most of our lands are more or less sloping—we expose soils to accelerated erosion by water or by wind and sometimes by both water and wind.
>
> In doing this we enter upon a regime of self-destructive agriculture. The direful results of this suicidal agriculture have in the past been escaped by migration to new land or, where this was not feasible, by terracing slopes with rock walls as was done in ancient Phoenicia, Peru, and China. Escape to new land is no longer a way out.

Lowdermilk wrote these words circa 1945. At the end of his paper, he added a plea to future readers:

> If Moses had foreseen what suicidal agriculture would do to the land of the Holy Earth, might [he] not have been inspired to deliver another Commandment to establish man's relation to the earth and to complete man's trinity of responsibilities to his Creator, to his fellow men, and to the Holy Earth. . . .
>
> Thou shalt inherit the Holy Earth as a faithful steward, conserving its resources and productivity from generation to generation. Thou shalt safeguard thy fields from soil erosion, thy living waters from drying up, thy forests from desolation, and protect thy hills from overgrazing by thy herds, that thy descendants may have abundance forever. If any shall fail in this stewardship of the land, thy fruitful fields shall become sterile stony ground and wasting gullies, and thy descendants shall decrease and live in poverty or perish from off the face of the earth.

Standing in Iglesia Emmanuel in East Porterville, a small town in the dusty Central Valley that has run out of water, I ask the pastor, Roman Hernandez, what actions his community is taking to combat their dire situation. He says, "We are basically praying for rain. That's what we are doing. Praying for rain."

In 2017 Pastor Hernandez's prayers were finally answered when the torrential ocean-born rains came to Southern California. But with cracked, hardened, dry dirt, much of the water that fell on the ground turned into flooding and was swept away. As the summer sun returned, California found itself once again with depleted aquifers and water problems.

The lesson is simple. Unless we restore the health of our soils, not even the rain will save the desert.

CHAPTER 4

MEET THE REGENETARIANS

Farms. That's what used to be in Anaheim, California.

Buttressed by mountains to the north and hugged by the Santa Ana River to the south, the area was a floodplain and grazing ground for massive herds of elk and bison. The minerals flowing down from the mountains, the fertilizer from the animals, and the dampness of the river basin all contributed to the area's rich soils. That's probably why it was chosen by fifty German immigrant families to create what was once the largest and most successful vineyard in California. It's why fields of orange trees were later grown here. And in a roundabout way, it's even why the hybridized boysenberry was cultivated and popularized by Walter Knott, who founded the family-farm-turned-roadside-attraction Knott's Berry Farm.

In 1954 about 160 acres of those Anaheim farms were turned into a construction site. A year later, a place called Disneyland opened its doors to the public. Drawing around sixteen million visitors a year and pulling in an estimated $1.8 billion in income, Disneyland has become the equivalent of Anaheim's economic sun.[1] Feeling the gravitational pull of the "Magic Castle," a smattering of large corporations and even ministries have since made the area their home, thus contributing to the area's prowess.

Today everything in Anaheim has a Disneyesque feel. In other words, big, fanciful, and plastic. Even Knott's Berry Farm grew up to become its own larger-than-life theme park. It's the perfect backdrop and stage for the largest food exposition in North America.

IT'S GETTING REAL

If you've never seen the YouTube music video entitled "Whole Foods Parking Lot," it's worth the four minutes. (At least the 6 million people who've watched it so far think so.) It features Dave Wittman, a well-heeled Caucasian rapper who is having a difficult time parking his Prius and getting the organic ingredients he needs for dinner at one of America's largest Whole Foods stores. As Dave points out, "It's getting *real* in the Whole Foods Parking Lot!" This cat-video competitor makes an important point: today the supply of organic food is barely able to keep up with demand.

This is why I'm driving to Anaheim. To attend the Natural Products Expo West. The event is put on by the vaguely ecclesiastical-sounding New Hope Network, a group that, according to its website, has "solutions for the complete supply chain from manufacturers, retailers/distributors, service providers and ingredient suppliers" and whose mission it is to "grow healthy markets."

After the three-hour battle with morning traffic through Los Angeles down to Anaheim, my stomach is starting to rumble. Another half hour to park and another fifteen minutes walking to the convention center and I am now officially starving. Lucky for me, some nice-looking young women in what amount to green cheerleader outfits are parked next to the sidewalk giving out some kind of vegan wafer-type bars that reminded me of Kit Kats. Mmmmm. "Yes, thank you," I say, taking a handful. Chomping a few down and feeling my glycemic index spike, I am suddenly beset by thirst.

Again, as luck would have it, right across the street from Disney as the trickle of people turns into a throng trying to push its way into the convention center, a few hip-looking young guys are handing out coconut water. "Perfect," I think, grabbing a couple of bottles and washing down the rest of my dry but tasty snacks. I float toward the convention center on a sugar high.

Having forgotten to register until the last minute, I stand in all three lines for a badge, each of which are no less than half a mile long. Again, a stream of food items is offered up like appetizers at a cocktail party.

Complete with a band stage, outdoor booths, and endless branding, the expo spans the entire convention center and the grounds around it. It also wholly envelops the neighboring Marriott, including its convention halls, conference rooms, and meeting rooms. One might assume that this expo is to hype up "natural foods" to consumers with a casual interest, but that could not be further from the truth.

With seven thousand tickets costing around $1,000 apiece and booths fetching up to $100K, this exposition isn't about the end consumer, it is a business-to-business frenzy to sell products to bulk buyers including grocers and online sales channels. As evidenced by the mass of people here, the business of organic is blossoming.

The last decade has seen breakneck speed growth in the organic foods industry. According to the Organic Trade Association, "Consumer demand has grown in double-digits every year since the 1990s—and organic sales increased from $3.6 billion in 1997 to over $39 billion in 2014" to over $43 billion in 2015. (That's billion with a big B.) This is the kind of growth that investors dream about. The tenfold increase in demand for organic food may have to do with the fact that 84 percent of American consumers now purchase organic foods at some stage each year.

Oddly, this growth is occurring inside a food market that as a whole is stagnating. This is largely because food marketers have reached the point where it is increasingly harder to sell more food (more calories) to a group of people (mostly Americans) who are already overstuffing themselves. At least on the demand side, we've reached "peak food." This is why organics represent the most exciting thing to hit grocery stores since, well, sliced bread. It's not that organic foods are a growth industry—it is that they represent an explosive growth in a sea of market nongrowth.

The good news for the world of organics is that there's plenty of

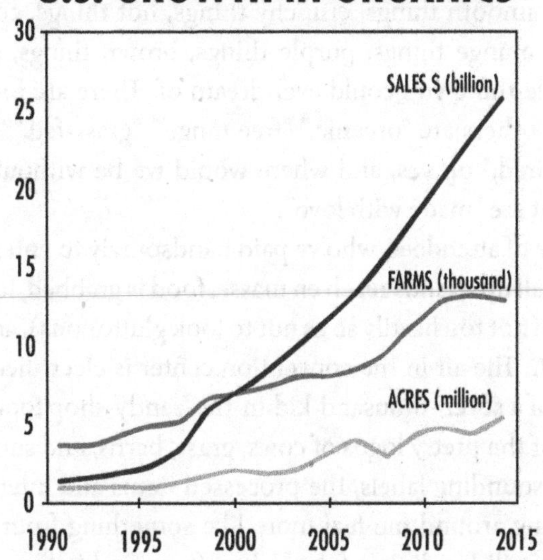

US demand for organic food is accelerating so quickly that it is outstripping supply, causing America to be a net importer of organic foods.

runway. Today, organics represent a tiny fraction of the total $5 trillion-plus food market in the United States. When it comes to organics, plants aren't the only thing awash in green. It's why even vaguely natural food companies are in Anaheim with their booths, tight-T-shirt-wearing pretty people, CEOs, and buffets of food samples.

Badge finally in hand, I walk past the branded banners and free-food purveyors and into something that almost defies description. Thousands of bodies move like tributaries through aisles filled with vibrant color and sumptuous smells. The sheer scale of it overwhelms my senses and makes me dizzy. The smiling booth models, with their trays, platters, and shelves of edible items, make me feel like I've entered some Tim Burton–*Alice in Wonderland*–type caloric universe.

There are pirates, unicorns, beach volleyball games complete with

VW buses and DJs, a life-size windmill, sweet things, salty things, fizzy things, smooth things, crunchy things, hot things, cold things, blue things, orange things, purple things, brown things, and every type of edible thing you could ever dream of. There are foods called "all natural," others are "organic," "free range," "grass-fed," "humane," "nature-inspired," oh yes, and where would we be without all those products that are "made with love"?

The surge of attendees, who've paid handsomely to enter, are literally eating it all up. Hands reach en masse, food is grabbed, it's inserted into mouths (not too hastily so as not to look gluttonous), and it starts all over again. The air in the convention center is electrified with the excitement of a seven-thousand-kid-in-the-candy-shop food orgy.

In spite of the pretty logos of cows, grass, barns, and sunrises, and all the lofty-sounding labels, the processed items and intensive marketing swirling around me feel more like something from a George Jetson take-a-pill-for-dinner future than from an idyllic wholesome "organic" farm. I see no corn on the cob, no lettuce, no zucchini. (Unless you count corn *chips*, premade salads in a bag, or dehydrated zucchini *crisps*.)

This "industrialized organic" world begs a deeper question. It's a question that I am just starting to remember is my reason for being here as I munch on another sample of an "all-natural" tamale and wash it down with a "healthy" red agave–sweetened carbonated beverage in a pleasingly slender glass bottle: *What the heck is "organic," really?*

THE WORD, THE MYTH, THE LEGEND

It's hard to argue that putting toxic chemicals into the food we eat and feed our children is a good idea. One can quibble about the effects of various amounts of this or that carcinogenic "residue" in our food, but spraying glyphosate, 2,4-D, Alar, daminozide, paraquat, endosulfan, aldicarb, methyl parathion, terbufox, barbofuran, methamidophos, methomyl, monocrotophos, chlorpyrifos, or ethoprophos, to name less than 1 percent of what the World Health Organization calls "extremely hazardous pesticides" onto things we eat fits the definition of *self-destructive, foolish,* and possibly even *masochistic.*

When it comes to industrial farming, we eaters, we human consumers of calories, have become among our own worst enemies. In contrast, "organic food" is food grown without these or other toxins.

The word *organic* comes from the Greek "organikos," meaning something relating to a body organ. That seems simple enough. But "organic" also refers to "organic chemistry" with its "organic compounds" and "organic molecules" (as opposed to "inorganic compounds"). The idea behind "organic" chemistry is that it can be found in living things. (Ironically, today most "organic molecules" can be synthesized in a lab through very nonnatural means.) Then there's all sorts of other uses of "organic," like the "organic matter" in soils or the "organic theory of the State" and the "organic theory of Society." One can see how all this could lead, "organically," to confusion.

As it relates to food, "certified organic" refers to a set of standards put forth by the US Department of Agriculture in 2002. A food passing these standards and being inspected on a regular basis is considered "USDA Organic." Kicking the habit of using lots of synthetic chemical inputs and making the transition to these rather strict organic standards is a challenge for farmers and ranchers who've been brought up and educated inside a system of agriculture that has reduced nature to a set of elementary chemistry equations.

To understand why these standards were developed and why they are important, we have to go back about a hundred years to the countryside of Germany and France.

As French minister of agriculture Stéphane Le Foll said, the farmers of Brittany recognized three critical aspects to agriculture: "manure, manure, and manure." Around the turn of the last century, this sentiment was echoed by a man named Rudolf Joseph Lorenz Steiner, who founded an esoteric movement called anthroposophy.

The central tenets of Steiner's school of thought pointed toward using the scientific method to better understand and work with some of the more mysterious aspects of our existence, specifically, spirituality, children, and our connection with the earth.

Steiner felt that as we grow up, humans leave the more "spiritual" connection we experience as children and move toward an "intellectualization" of the world and in so doing, we tend to lose our connection with nature. Steiner encouraged a philosophy of life that embraces both childlike wonderment and scientific thought. This is why Rudolf Steiner developed Waldorf schools. He was a big proponent of people finding their own spiritual freedom, rather than learning it from an authority. He was also big on something called "organic" agriculture.

Through a series of eight lectures that he delivered in Germany in 1924, Steiner defined the first form of modern organic agriculture. He captured a small but growing zeitgeist in Europe that agriculture could be done at a substantial but localized scale without the use of synthetic chemicals or fertilizers. To do so, he asserted, one had to understand the cycles of nature. The codification of this idea was a reaction to the pressure to create a centralized, simplified, and stable food supply, especially for Germany.

By the end of the First World War, Europe had surged toward industrialization. To feed its growing population, Germany needed more nitrogen for its rather poor soils. The country went from using bird guano from Peru to saltpeter (potassium nitrate) from the Peruvian desert to calcium cyanamide, which it had to manufacture, and then finally synthetic ammonium nitrate made from fossil fuel through the Haber–Bosch Process.

A number of traditional farmers who had been working the land handed down to them through the centuries felt these industrial measures were extreme and unnecessary. Their experience showed them that vast amounts of food could be grown without "external" inputs. Steiner coalesced their anti–industrial farming sentiment into a form of agriculture he called "biodynamic" farming (*bio* from the Greek "bios," meaning "living," and *dynamic* from the Greek "dunamis," meaning "power"). In other words, a form of agriculture that employs the power of life to create life. This flew (and still flies) in the face of the more intellectualized model of inject, manufacture, and extract calories—the same model we now accept as "normal" agriculture.

Steiner said that the animals, crops, and soil on a farm are really a single system. He encouraged the use of herbal and mineral elements as well as compost and manure to augment soil health. He emphasized that cows are a critical component of the growing of food. Like the old French farmers, the Phoenicians who preceded them, and the Mesopotamians who brought cows from the Fertile Crescent, he believed that without the manure from bovines, it is impossible to maintain healthy soil. He believed that planting according to a lunar calendar was important because it affects the response of plants, pests, and the soil. While some of his ideas were pure mysticism, time and science would prove that much of his rationale stood on solid ground.

In 1993 as a long-haired youth, I joined an international farming volunteer organization and headed off to Europe to work on small organic farms. I can testify that Steiner's lectures brought forth methods that have been dutifully practiced for the better part of the last century, all the while building soil health. There are many small-scale farmers in France and Germany who have been farming biodynamically for three or four generations.

Farming strictly according to Steiner's guidelines is hard, manual labor–intensive work. But on the farms on which I worked, the soil is rich, the animals are raised in a healthy, humane, and integrated way, and most importantly, the per acre food production is through the roof. Some believe the three hundred thousand or so acres of biodynamic farming that are practiced today are simply occult pseudoscience, but a number of studies point to the benefits of biodynamic farming.

While the mystical aspects of Steiner's philosophy will continue to cast doubt on the validity of biodynamic farming as a superior agricultural model, Steiner unarguably contributed something important—a delineation between farming that is done through industrialized external inputs (which are themselves based on limited resources) versus farming that is done by working within the confines of a natural system. In many ways, this critical delineation makes him the great-grandfather of organic agriculture.

AMERICA'S ORGANIC ROOTS

Getting to the Rodale Institute in Pennsylvania from the hustle and bustle of New York City is a visceral transition. Leaving the cement, brick, and steel one travels through the industrial world of "Jersey" and past some of the decrepit towns that were once part of America's industrial might. On a lark, I pull up Billy Joel's "Allentown" on my iPod. It's been a while since I've heard him warble about America's crumbling coal and factory towns. His Rust Belt rock fits with the scenery. And whaddyaknow, there's Bruce Springsteen's "Born in the USA" up next on the playlist.

Eventually greener pastures replace crumbling edifices. Kutztown, PA, has a small college that breathes some life into the town in the form of a coffee shop, a microbrewery, and, well, college students. That works out for the Rodale Institute, which is just a few miles out of town and has numerous farm experiments and operations. A steady stream of interns is always at the ready to help.

The grounds of Rodale are picturesque Pennsylvania. Whitewashed wooden buildings complement a few old stone storehouses and, along with a huge red barn, the place looks like the physical embodiment of Old MacDonald's farm. I get a tour from a bearded chipper young lad who drives me around in a golf cart. Compost piles, check. Bees, check. Organic apple orchard, check. Cage-free chickens, check. Happy-looking pigs grunting in and out of the uber-clean pig house, check. Fields of growing food, check. The place is 333 acres of veritable organic farming perfection. But it didn't start that way.

In 1920, Jerome Cohen, a young Jewish tax auditor who had grown up amid harsh anti-Semitic prejudice in Manhattan, decided to change his name to Jerome Rodale. For the next twenty years, Jerome built a manufacturing business with his brothers, created a family with his wife, Anna, and began a modest publishing company. In 1940 Jerome and Anna Rodale purchased a run-down farmhouse on a tired sixty-three-acre farm in Emmaus, Pennsylvania. This old farm changed the course of their lives and in so doing sparked a movement that may one day change what we eat. I'm talking about that word again, *organic*.

Rodale was inspired by an Englishman named Sir Albert Howard, who had worked as the Imperial Economic Botanist to the government of India. Howard was fascinated by the difference between soil health in the West and East. He said that in Asia the "balance between livestock and crops is always maintained" and "the crops are able to withstand the inroads of insects and fungi without a thin film of protective poison." He compared this with his experience of agriculture in Western countries in which, he said, "Agriculture has become unbalanced: the land is in revolt: diseases of all kinds are on the increase."

The difference, Howard postulated, was that farmers in the East were using a delicate balance of animals and land care to build soil *humus*. He defined *humus* as the "complex residue of partly oxidized vegetable and animal matter together with the substances synthesized by the fungi and bacteria which break down these wastes." The more humus, the better the health of the soil and the healthier the crops. Today humus is often referred to as soil organic matter (SOM) or just organic matter, for short.

As the United States lurched from the widespread hunger of the Great Depression to the destruction of farmland during the Dust Bowl, Jerome Rodale saw Howard's ideas as critical and prescient. Rodale and his family began to experiment with organic agriculture on their farm. Rodale began to publish a circular called *Organic Gardening and Farming*. At first the magazine was a failure. Rodale later renamed it *Organic Gardening*. In 1945 he wrote *Pay Dirt*, a name apropos of his successful farming venture, which by then was growing copious amounts of food without the use of pesticides or chemicals. The book was well reviewed, sold tens of thousands of copies, and pushed the name J. I. Rodale into the public sphere as "Mr. Organic."

Pay Dirt did something else too. It fanned the flames of J. I. Rodale's passion to catalyze a new movement. The United States was emerging from its biggest experiment with industrial machines—World War II. The country was ready to put the war machine to work at home to hypermechanize food production. In some ways, Rodale had gotten his foot in the door just before organic was firmly shut out. Organic may not have been big, but because of the growing popularity of Rodale's magazines and books, it wasn't going away.

By 1970, when buxom blond model and health food store owner Gunilla Knutsson was splashed across the cover of *Life* holding a cornucopia of fresh vegetables with the subhead "ORGANIC FOOD: NEW AND NATURAL," organic was clearly growing from a counterculture experiment into a consumable idea. The Rodale family would continue their work by separating their farming enterprise and naming it the Rodale Institute. On the publishing side, their magazines include *Prevention* and *Runner's World*. On the farming side, their family made another important distinction. They began a side-by-side comparison of nonirrigated organic versus conventional farming techniques.

The Rodale Institute's Farming Systems Trial has been going on for more than three decades. It is the longest-running such comparison in the United States and one of the longest-running comparisons in the world. To summarize the findings of their thirty-year report: overall, organic produces the same or more quantity of food per acre than conventional "industrialized agriculture." There may be some seasons that produce slightly less, but on balance the production is very similar. Even with this data in hand, many people familiar with farming are quick to point out "Yes, but organic is far more labor-intensive."

While this may be true it misses a critical point. The principal reason organic farming may be "more difficult" by today's standards is that the vast majority of infrastructure that supports farming and food manufacturing in the United States is built on the post–World War II model of subjugating nature with petroleum and machines. If we put aside for a moment the idea that dominating nature by force is the correct approach, a new question emerges: "If we are not gaining productivity, then what exactly do we gain from all those expensive chemicals, pesticides, fertilizers, and sprays?" In other words, if organic agriculture and conventional agriculture produce approximately the same yields, then what is the point of the "modern industrial agriculture system?" Is it "efficiency"?

If so, would it not be most efficient to simply employ "the power of life to create life," as Steiner had preached? This is precisely why, despite an entire modern agricultural system set up to enrich corporate profits (not the soil), USDA Certified Organic farms and the food they produce are spreading like weeds.

As movements tend to do, the organic food movement is morphing as rapidly as it is growing. On one hand, the Natural Products Expo and the success of the Whole Foods grocery chain show that organic food is evolving into mass-processing and commercialization. But at the same time, it is also moving in another, less publicized and far more important direction.

This direction involves using agriculture not just to produce food, but also to manage the restorative capacity of nature. To get it right, this system of agriculture has undergone a lot of experimentation. And some of that experimentation was a terrible failure.

THE ELEPHANT IN THE ROOM

Say the name Allan Savory in a room of foodies, health nuts, or people who avidly read about farming and you're bound to get mixed reactions. I made this mistake at a family gathering. A relative of mine who works in the dairy industry and happens to be vegetarian overheard me mention that I was planning to interview Allan Savory. His reaction was quick and fierce. "That guy? I *HATE* that guy!" he said. "He is the biggest *shyster* ever to walk the planet!" His anger left me backpedaling.

As I write these words, Allan Savory, who is originally from Zimbabwe, is in his early eighties. His TED Talk alone has well over three million views. His Savory Institute has thirty "Savory Hubs" around the world that manage hundreds of millions of acres of land. His books on what he calls "rangeland management" have influenced an entirely new way of thinking around animal grazing patterns.

I meet Savory one evening after a talk he is giving in Los Angeles. Having heard the legends of this man, I expect a towering giant of an

African ranger complete with one of those British adventure outfits. In the flesh, Savory couldn't be more the opposite. He is a diminutive, soft-spoken, gentle man with eyes that sparkle like a fox full of mischief.

To be sure, Savory inspires vitriol from some and devotion from others. To understand why, I ask him why he killed forty thousand elephants.

In 1965 Savory, who was at the time a young zoologist fresh out of university, joined the Game and Wildlife Department of his native Zimbabwe (then known as Rhodesia). According to Savory, "I began to see massive environmental degradation that was a threat to the wildlife. And although our job was mainly to just burn the land, get the grass flushing green for all the wildlife, and stop the poachers, I could see the damage it was doing, and in fact, I became somewhat unpopular by saying I thought we bureaucrats in the Game Department were a greater danger to the wildlife than the poachers were." From there, Savory would make his opinions louder and his popularity would continue to decline.

Savory was committed to finding a way to stop the degradation of the lands his department was managing. The Zimbabwe Game Department had removed all the native humans from their most picturesque lands in an effort to create a series of national parks. With the tribespeople gone, the degradation of land worsened. Savory explains that at the time, nobody realized it was the humans that, through their drumbeating at night and hunting during the day, kept the herd animals moving throughout these delicate ecosystems, ensuring those animals never stayed in one place too long.

Savory surveyed the research on the subject of land degradation, he questioned the experts, and he came up with the only seemingly viable plan he could find. Says Savory: "Now the only belief of scientists worldwide is if you've got land deterioration like that, it is caused by too many animals. And we didn't have any livestock, so it had to be too many elephants, buffalo, and so on. So then it was easy for me to gather the data." Savory led an entire committee of scientists who concluded that the national parks they were managing had fallen prey to the problem of "overgrazing" by wildlife. It seemed that the only solution was to remove a vast number of the most destructive of those creatures—the elephants.

Savory says, "Only when we went ahead and shot forty thousand elephants and it got worse, did I say 'Okay, something is very very wrong here.' Because the result was the opposite of my research. What I had done was twist my data to fit the belief. And that's what American scientists have been doing for a hundred years. We were all wrong. There weren't too many animals—there were too few." In spite of the fact that he insists he loves elephants, Allan Savory would forever be branded "the elephant killer."

After his failure with culling wildlife, Savory tried going in the opposite direction by helping to create Africa's "game ranching industry." He explains that "Our belief was, if we could just get rid of all the livestock, cattle, goats, sheep, et cetera and just manage wildlife, we could reverse the land degradation. Well, again we were wrong. We had changed the behavior of the wildlife so much that we couldn't do it." In other words, reintroducing the native animals back onto the land did not work either.

Allan Savory began to think there was yet another approach that didn't involve killing wild animals or trying to manage them. He explains, "I realized that the healthiest land I'd seen was where there were the highest numbers of elephant and buffalo et cetera, but they weren't lingering. So that was in my mind constantly. I then started to think that we had to use livestock." Savory says that because relationships between animals and the land have been so altered by human intervention, it is nearly impossible to go backward and reinitiate migratory patterns of large herd beasts across vast swaths of range, swaths that rarely even exist anymore.

"At one point I was looking at livestock, to see what might happen there, and there was a place in a corner of a fence where a big flock of sheep had concentrated during rain, and it was beautiful. And the farmers that were with me couldn't understand; I just got down on my hands and knees, burying my fingers in the soil. And they said 'What are you so excited about?' and I said, 'This is it! I've suddenly seen that livestock can do what wildlife do.' "

That epiphany was the basis for the rest of Savory's career and the development of his principles of rangeland ecology and rangeland management. "And that led us to where we are today, to realizing that only livestock can do it [restore soil health] in practical terms, because we've lost most of the predators. It was the predators that regulated the behavior of the big herding animals," he says.

In lieu of predators (including humans), which served as the impetus to move huge herds of herbivores across the open spaces on Planet Earth, Savory saw that soil health could be restored using strategic grazing. By keeping a managed herd constantly on the move, Savory believes we can use the disturbance caused by their hooves and the fertility bestowed by their manure as two critical tools in the battle to restore the health of Earth's soils.

While the name Allan Savory will likely always cause controversy, his failures set the stage for an undeniable truth. To build and maintain its soils, nature uses both plants and animals in a dance of give-and-take. Removing either half of the equation (the plants or the animals) is certain to shortchange the very foundation for all of the aboveground life—the soil. Savory contributed this piece of the puzzle, but many more ideas would have to fall into place in order to synthesize the viable, replicable, and scalable model of farming and ranching called regenerative agriculture.

TOWARD A PERMANENT AGRICULTURE

In the 1970s "natural food stores" appeared in places like Berkeley, California, and "food co-ops" began in a handful of big cities across the United States. By the mid-1970s a book by Japanese farmer Masanobu Fukuoka called *The One-Straw Revolution* had reached the English-speaking world. Fukuoka's experience with fruit trees and rice farming taught him that practices like tilling the soil and heavily pruning trees, which were thought to enhance agricultural productivity, often decreased it.

Inspired by Fukuoka, in Australia, a graduate student named David

Holmgren and his professor Bill Mollison coined the term *permaculture,* referring to a type of agriculture they believed could sustain a "permanent agriculture." According to Mollison, "Permaculture is a philosophy of working with, rather than against nature; of protracted and thoughtful observation rather than protracted and thoughtless labor; and of looking at plants and animals in all their functions, rather than treating any area as a single-product system." By "single-product," Mollison is speaking about chemical-industrial agriculture's model of producing, for example, only corn or soy.

In a permaculture system, the home and/or farm buildings are placed at the center of a system of "zones." Moving outward from the domicile, one moves from delicate plants and operations (strawberries and worm composting, for example) toward more robust plants (trees) and eventually into a forested wilderness-type barrier. The growing of crops and grazing of animals occur in the zones that correspond to their need for attention and their effect on the other zones. Permaculture encourages the use of agroforestry (using trees with crops), natural building (using things from nature to build fences, etc.), rainwater harvesting, mulching, and managed intensive rotational grazing (MIRG).

Regardless of geography, overgrazing invariably degrades land. Too many animals eating too much forage in too small a space causes soil to become bare and exposed. Left unattended that soil loses its moisture, becomes brittle, and is carried away by wind and rain. As root structures disintegrate, plant life has a more difficult time reestablishing itself. Dirt begets more dirt. Desert begets more desert. The animals are moved to a new location and the cycle begins again, marching the frontier of dirt ever forward as the cover of green retreats.

In contrast, managed intensive rotational grazing, also known as cell grazing, mob grazing, or even eco-grazing, is modeled after the behavior of herds in nature. The idea is to move a herd of ruminants or nonruminants according to a balance of how much they have eaten and how much they have fertilized. In order to gauge how long the herd should remain in one place before being moved, MIRG requires an acute awareness of the grasses, the watershed, and the ecosystem

Permaculture offers a way to begin working toward a "permanent agriculture" that could sustain humanity well into the future.

in which the herd is foraging. The herd's disturbance of an area is followed by a period of rest that allows new pasture, range, or forest growth. The deposits of fertilizer (manure), the height of the grass after eating, and the level of hoof disturbance from the animals are all taken into consideration. When used judiciously, these impacts can spur new growth.

Perhaps nobody has done more to popularize MIRG than Joel Salatin, whose Polyface Farm is featured in the popular documentary *Food, Inc.* and in Michael Pollan's revelatory book *The Omnivore's Dilemma*. Salatin's unique views are also expressed in his books *Everything I Want to Do Is Illegal* and *Folks, This Ain't Normal: A Farmer's Ad-*

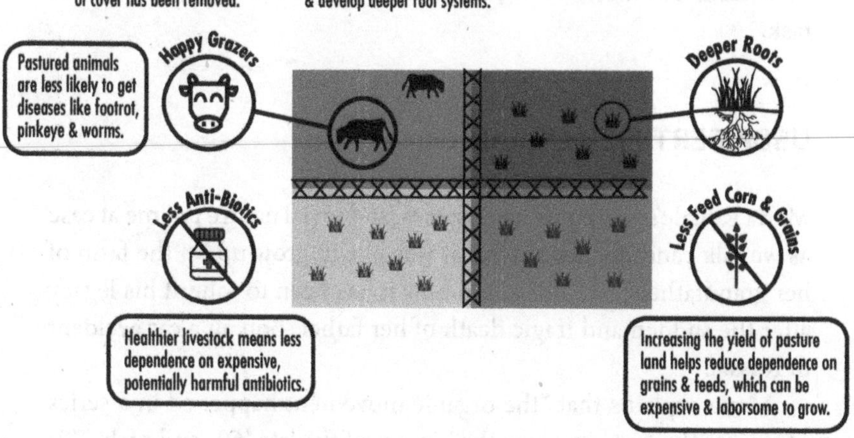
WHAT IS ROTATIONAL GRAZING?
Large pastures are broken into sections called paddocks. Livestock are moved between paddocks when 50% to 70% of cover has been removed.
Paddocks not currently being grazed are left to rest for a period of 25–30 days. This allows plant life to come back to grazing height & develop deeper root systems.
Deeper root systems allow plants to draw more nutrients from the soil. They also help reduce field erosion which can lead to desertification.
Pastured animals are less likely to get diseases like footrot, pinkeye & worms.
Happy Grazers
Deeper Roots
Less Anti-Biotics
Less Feed Corn & Grains
Healthier livestock means less dependence on expensive, potentially harmful antibiotics.
Increasing the yield of pasture land helps reduce dependence on grains & feeds, which can be expensive & laborsome to grow.

vice for Happier Hens, Healthier People, and a Better World. Like many farmers who practice some aspects of permaculture, Salatin maintains he is not in fact a "permaculturalist" but rather a "grass farmer."

Probably more than any other system of thinking around agriculture, permaculture has firmly grounded the growing and consuming of organic food into a type of social activism. After all, it's one thing to tell people that tomatoes without pesticides are better for them, it's another to say that growing food in relative harmony with the ecosystem is an economic, social, ecological, and even a political act.

This mixture of defiance against a calorie-producing system that is inherently destructive to people, plants, animals, and the planet while simultaneously participating in what is ideally a tangible, localized, and healthier cycle of food growth, consumption, and composting is scintillatingly attractive. While permaculture may have codified it, the organic movement has its roots firmly planted in the ideals of self-reliance, community development, anticorporatism, and freedom of choice—lofty ideals indeed. And while they have been expressed in various systems from biodynamic farming to rangeland ecology to organic to permaculture, without a set of agreed-upon standards for organic agriculture none of these ideas could truly scale up.

It turns out that creating those standards was a nearly impossible task.

USDA CERTIFIED ORGANIC

Maria Rodale's wavy gray hair, glasses, and jovial nature put me at ease as we talk candidly about what it was like to grow up on the farm of her grandfather J. I. Rodale and how it has been to inherit his legacy after the sudden and tragic death of her father, Bob, in a car accident in Russia.

Maria explains that "the organic movement happened in a series of waves. The first wave was the hippies of the late '60s and early '70s and for them it was about health." Of course, it was also political, especially on the West Coast, where food co-ops, health food stores, and macrobiotic restaurants were seen as part of a movement away from corporatocracy. She says that "in the 1980s the movement was growing in terms of the desire for certification. There were a lot of state certifications but the different states had different definitions. Finally, the process of USDA certification started around 1990."

Long before to the USDA's adoption of the organic standard, Bob Rodale initiated the Farming Systems Trial, the side-by-side comparison of organic versus conventional crops that continues to this day. The trial gave the organic movement something it had been missing—a scientifically verifiable, long-term study of the viability of organic farming.

Meanwhile, the USDA set out to do its own study. In 1980 they published the *Report and Recommendations on Organic Farming*. The USDA team interviewed a large number of organic farmers across the United States and Europe. Ultimately, their report was positive and pointed to the environmental benefits of organic farming, its judicious

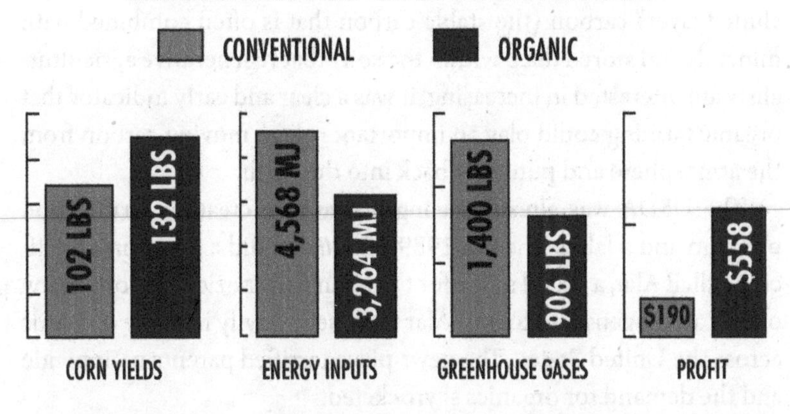

The Rodale Farming Systems Trail has been going for over thirty years. It shows that, overall, on a per-acre basis in nonirrigated systems, organic outperforms conventional agriculture.

use of resources, and its innovative approaches to disease and pest management.

Over at the Center for the Biology of Natural Systems (CBNS) at Washington University in St. Louis yet another organic study was under way. The study centered on organic agriculture in the Corn Belt of the Midwest. It found that organic farms growing corn, soy, and small grains consumed 60 percent less energy per dollar of crops produced. They also had 33 percent less soil erosion than conventional farms. While their yields were slightly lower and labor was more intensive, this was completely offset by the radical reduction on cost of inputs.

In a time before organics were popular, before standards were recognized nationally (never mind internationally), organic farms in the Midwest were, on a dollar-for-dollar basis, making more money than conventional farms in the same area. This was also well before "organic" produce fetched a price premium.

These farms also did something else that at the time was not earth-shaking but in retrospect is important. They sequestered more carbon in the soil. Soil organic matter (SOM) is 50 percent carbon by weight, and these first large-scale organic farms were keeping more of that in the ground. While this isn't necessarily the semipermanent "occluded layer" carbon (the stable carbon that is often combined with minerals and stored deep within the soil) that regenerative agriculturalists are interested in increasing, it was a clear and early indicator that organic farming could play an important role in moving carbon from the atmosphere and putting it back into the earth.

The USDA was already feeling pressure to create a certification program and a label. Then in 1989 *60 Minutes* did a piece on a pesticide called Alar, a brand name for the toxin daminozide. Among many other carcinogens and toxins, Alar was used heavily in apple orchards across the United States. The news piece terrified parents nationwide and the demand for organics skyrocketed.

Unscrupulous marketers began slapping "organic" labels on everything. It was high time to do something to certify organics. Representatives of the organic community headed to Washington, DC. Less than a year later, the Organic Foods Production Act of 1990 was passed by Congress.

The act called for a National Organic Program (NOP) and facilitated a fifteen-person board to draft the organic standards. The board had open public meetings several times a year in various parts of the country to garner input. In 1997, when the NOP finally published the draft of organic standards, it was clear they had ignored most of the input they had received. The draft of standards was riddled with ideas counter to the practice of organic farming. These included the allowance of food irradiation, sewage sludge as fertilizer, and the use of genetically engineered crops and other genetically modified organisms (GMOs).

The outrage from the organic community was tremendous. Two hundred seventy-five thousand public comments flowed into the USDA from concerned citizens (the most comments ever received by the USDA on any issue). Considering that there were only five thousand organic farms at the time the response was impressive. The USDA had inadvertently coalesced the organic movement that J. I.

Rodale had dedicated the better half of his life to. And collectively this movement was a leviathan. Awake, aware, organized, and enraged, the organic movement was ready to fight for a true organic standard.

The NOP took three more years and in 2000 published a far less controversial and more acceptable draft of the organic standards. Another round of public comments followed and in 2002 the Final Rule of the National Organic Standard went into full effect. Fifty-seven years after the publication of his book *Pay Dirt: Farming and Gardening with Composts*, J. I. Rodale's vision had come to life. With a set of clear standards in place and an organic food movement of unprecedented proportions supporting them, the world of certified organic food was set to grow its way into the mainstream.

A GOLD STANDARD

Apart from kosher foods, USDA Certified Organic represents the strictest set of food standards in America and one of the strictest sets of standards in the world. But before we go into the standard itself, there is an important distinction that bears mention.

"All natural," "free range," "made with love," and the other labels I referenced earlier do not have a certification program. There is no set of standards for them. "All natural" applies as equally to motor oil as it does to potatoes. "Free range" can mean that chickens are grown in factory farmhouses with locked doors.

The USDA Organic label is not perfect by any stretch of the imagination, but it is the only organic food label that is backed by inspectors, standards, and a strict government program. It's also the only label that took over sixty years of gnashing of teeth and constant work on the behalf of organic farmers to create.

It is also the only government-backed standard that lays the groundwork for regenerative agriculture.

To receive the USDA Organic label, a farmer must not use genetically modified organisms (GMOs), must not use sewage sludge on his or her fields, and must not feed his or her livestock

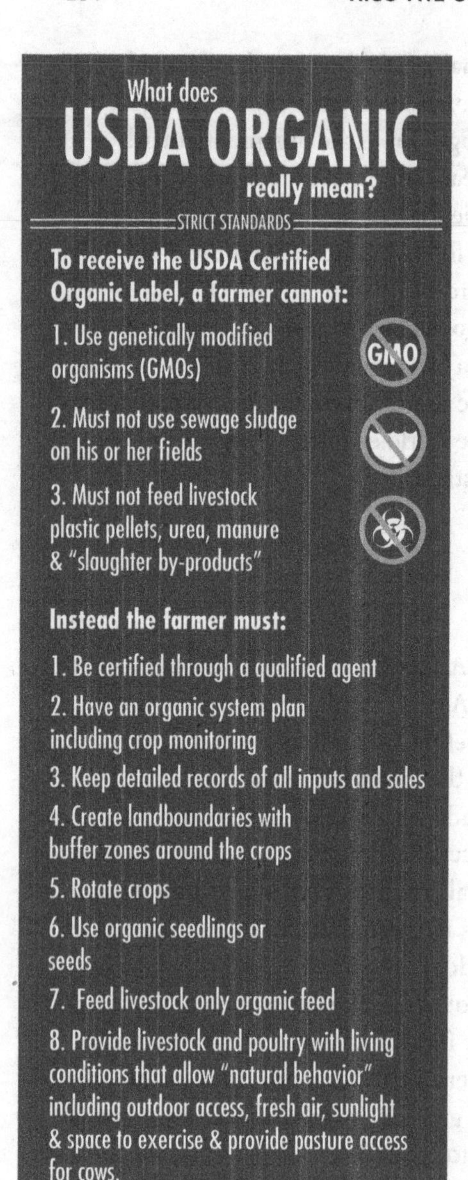
What does
USDA ORGANIC
really mean?
STRICT STANDARDS
To receive the USDA Certified Organic Label, a farmer cannot:
1. Use genetically modified organisms (GMOs)
GMO
2. Must not use sewage sludge on his or her fields
3. Must not feed livestock plastic pellets, urea, manure & "slaughter by-products"
Instead the farmer must:
1. Be certified through a qualified agent
2. Have an organic system plan including crop monitoring
3. Keep detailed records of all inputs and sales
4. Create landboundaries with buffer zones around the crops
5. Rotate crops
6. Use organic seedlings or seeds
7. Feed livestock only organic feed
8. Provide livestock and poultry with living conditions that allow "natural behavior" including outdoor access, fresh air, sunlight & space to exercise & provide pasture access for cows.

plastic pellets, urea, manure, and "slaughter by-products."[2] Instead the farmer must: be certified through a qualified agent, have an organic system plan including crop monitoring, keep detailed records of all inputs and sales, create land boundaries with buffer zones around the crops, rotate crops, use organic seedlings or seeds, feed livestock only organic feed, provide livestock and poultry with living conditions that allow "natural behavior," including outdoor access, fresh air, sunlight, and space to exercise, and provide pasture access for cows.

What is remarkable is that this list applies only to a small portion of the food grown in America today. The vast majority of calories produced have none of these stipulations. Thus, "food security" and "food safety," two concepts that seem to go hand in hand, are very much in opposition in modern agriculture.

There's a push to make the practice of organic farming better, so much better in fact that "certified organic" may one day mean that a food isn't just free from hormones and chemicals, but that it's actually healing the soil and reversing climate change.

To do that, however, requires boldly going where nobody has gone before. I'm not talking about going up to the stars. I'm talking about going downward into the earth and into a universe so infinite that it makes the vastness of space seem trivial.

DEEP EARTH LABORATORY

Dr. Kristine Nichols meets me in the old white wooden framed house that serves as Rodale's headquarters wearing a shirt that simply says "KALE." Her hair is cut short and her glasses seem like an ever-present feature that frame a pair of inquisitive eyes. Eyes, which I learn, that have spent the better part of three decades peering into a universe few have seen.

Kris, as she likes to be called, grew up on a Midwest farm that her father purchased the year she was born. "As many teenagers do, I wanted to run as far and as fast away from what my family was doing," she says as we walk across the gravel parking lot toward an unassuming white building. "And running away from agriculture was where I wanted to go."

We walk through a musty storeroom filled with long clear cylindrical "soil core" samples and into a laboratory. As an undergraduate student Kris ended up working in a research lab not unlike the one we're standing in. That's where she first came face-to-face with a particular family of fungus called mycorrhizal fungi. There are few people on Earth who have spent more time learning about mycorrhizal fungi than Dr. Kris Nichols. One could say it's a bit of an obsession.

We walk upstairs to her "clean room" laboratory and grab a couple of stools as she begins to explain why she has dedicated her life to looking down. "The largest organisms in the world are soil-born fungi," she says. "These are microscopic, but they can grow into these very large networks. The largest ones have existed for more than 2,500 years. They are mostly in forest ecosystems, and in fact the largest one is in the Pacific Northwest." This particular fungus, an *Armillaria ostoyae* in eastern Oregon, stretches out over 2,400 acres. While a

2,400-acre fungus may sound like a housecleaner's worst nightmare, it may actually be humanity's best friend.

Asked why a fungus, albeit a large one, has consumed so much of her life, Kris explains it thusly: "These mycorrhizal fungi are the key to life as we know it. Human life would not exist without mycorrhizal fungi. Plant life would not exist without mycorrhizal fungi. We would still be in an aquatic environment without mycorrhizal fungi. It doesn't mean there wouldn't be a Planet Earth, it just means that *we* wouldn't be here."

She says that through RNA analysis scientists have been able to determine that an ancient ancestor to that 2,400-acre fungal network was the thing that helped algae move from pools of murky water onto land.

Kris gives me a brief lesson in evolutionary biology. "Pretty much all life on Planet Earth is carbon based. And every cell, every molecule, has carbon building blocks." All that carbon gets into plants via fungi. "For plants, the root structure was an anchoring structure. It was a foot into that mineral environment, so that way if a plant grew vertical, it wouldn't fall over. So because roots weren't designed initially to be absorptive structures they became reliant on the fungus to do this," she explains. Apparently the fossilized record shows mycorrhizal fungi inhabiting the roots of plants as far back as 500 million years.

Apart from making all life (including human life) possible, mycorrhizal fungi did something else important. "These fungi and their relationships with bacteria and other soil microorganisms are what created soil," says Kris, continuing with her lesson. "These fungi are the things that can transport and provide for up to 90 percent of the nitrogen and phosphorus needs of a crop plant." I see why this might be an important, and grossly underrepresented, field of study.

I ask Kris if mycorrhizal fungi affect us personally. I mean, is this all about plants, or do we ever-important humans have a direct relationship with these little critters? Her answer makes me a little squeamish. "We are walking bags of bacteria. You have more bacterial cells in your body than you have human cells." (I must have been asleep when this was covered in biology class.)

Not noticing my paleness, Kris smiles with the analogy she is about to make. "So when you eat kale" (remember her KALE shirt?) "those nutrients in kale can't be absorbed into our bloodstream without their being broken down by the bacteria that are in our guts. When you are eating kale, your body doesn't consume kale. The bacteria consume the kale and you feed off what the bacteria have processed and released by the consumption of that kale. This is what's happening in the soil."

Since we humans are basically eating bacteria poop, the question becomes who is serving whom. Are we really just smart monkeys whose bodies are here to enable a five-hundred-million-year-old life-form, or is it the other way around?

Regardless of the answer, these small life-forms vastly outnumber us. Says Kris: "In every handful of healthy soil, there are more organisms than the number of people who have ever lived on Planet Earth. We are talking about well over 10 billion organisms that can be in a handful of healthy soil. Those organisms are processing organic matter that is in the soil. They are processing minerals and breaking those down and putting the nutrients that are there into a form that the plant needs."

Bad news. Not only are we humans the minority but a fungus determines whether we live or die.

When I think of fungus, I think of something that grows on old oranges or stuff my wife likes to leave in the fridge as her own biology experiments. But Kris says mycorrhizal fungi look very different from the fuzzy peaches in our chiller. She says, "The bodies of most fungi are these fine threads. Under a microscope, the roots look like tufts of cotton. And usually, what you see is layers of hundreds of individual threads of fungal hyphae even though it looks like one single thread."

Indeed, through a microscope, these mycorrhizal fungi look like the roots of the roots—small fibrous hairlike structures that seem infinite, delicate, and far more beautiful than gross. This is why some people call mycorrhizal fungi "the Internet of the soil"—they literally look like a rendering of World Wide Web connections. It makes

Mycorrhizal fungi are called the "Internet of the soil" because they literally connect millions of other species through their long, tentacle-like hyphae.
Main photo: Professor David Johnson, University of Manchester. *Magnified image:* Jeff Anderson, www.mycorrhizae.com.

one wonder if our highest tech isn't just replicating what has been in the soil for millennia.

Seeing as this underground networked life-form is the basis for so much life aboveground, I ask Kris why we don't hear more about it, and why I didn't learn all this stuff in Biology 101. She says that when she was an undergrad, the assumption was science had identified about 10 percent of what lives inside the soil. With increased resolution microscopes and better science, she now believes we can identify 0.1 percent of the life underground. "I tell my dad he paid for my college so I could become one hundred times stupider," she jokes.

She continues: "People have invested their entire careers on this and we are still struggling with trying to figure this out because of the complexity and the number of different chemical reactions and signals that can happen. There are roughly about ten million different carbon compounds that can be formed. Each of those can trigger something different in this environment. It's like a ten-billion-to-ten-million factorial. We have roughly thirty nonillion different types of bacteria. Nonillion is ten to the thirtieth power."

I can't even begin to fathom these numbers. For a quick comparison, think about the number of stars in our galaxy, the Milky Way. It's somewhere in the range of one hundred billion to four hundred billion. That means if you dig your hand into some rich soil and pull up a handful, you hold vastly more interactions of life than there are stars in our galaxy.

In that handful of rich soil, you hold an entire cosmos.

KILLING THEM SOFTLY

So mycorrhizal fungi are cool. But what do they have to do with the way our food is produced?

"Are we killing them with the way we do agriculture?" I ask.

"It doesn't mean today that all our soils that are under chemical conventional agriculture are completely devoid of microorganisms," Kris says. "The problem is that there is very little diversity in those microorganisms. You may still have mycorrhizal fungi, but they don't have anything to transport now so the system starts to break apart. And there is also a lot more disease pressure." The problem isn't just with the practice of conventional agriculture; it's with the entire mind-set.

"The problem is, we teach agriculture from a chemical perspective. We treat the soil as a bench top. And that's not how nature works. When you go out into a field, you have biology," explains Kris. The idea she (like many "soilophiles") is getting at here is that we are using old science (chemistry) to address a relatively new understanding of food production (which is all biological). In other words, we're trying to World War II our way to better food.

Nowhere is that more clear than with our use of copious amounts of synthetic nitrogen fertilizer. Kris says that in terms of nitrogen, "roughly 50 percent will end up in your crop plant. I don't know many people who would buy a product that is 50 percent deficient, but that's essentially what we are doing in conventional agriculture with nitrogen fertilizer. With phosphorus fertilizer it's even worse. Our nutrient-use efficiency of phosphorus is about 30 percent." The reason these chemical additives are so ineffective, she explains, is because they try to deliver nutrients into the roots rather than addressing what the mycorrhizal fungi need.

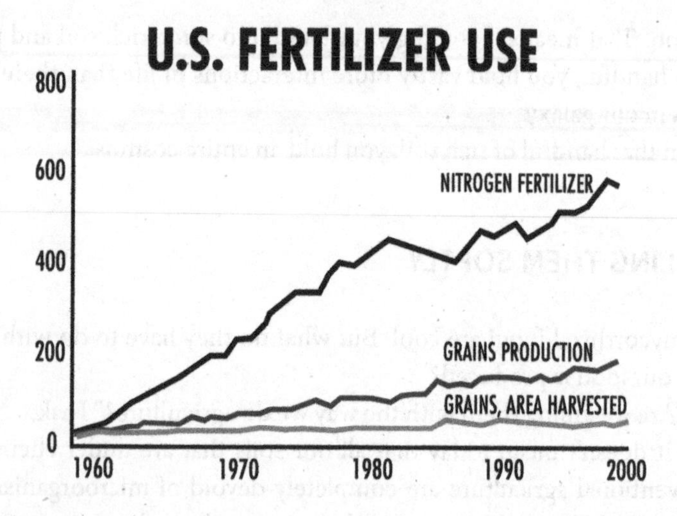

Today our overproduction of grains is possible only through precipitously increased use of synthetic nitrogen.

"On a per-bushel basis our nutrient-use efficiency, especially nitrogen, has declined. So although we have increased the amount of crops we are producing per acre, it's actually taking more nitrogen on a per-bushel basis to produce a bushel of grain today than it took to produce a bushel of grain in 1960," says Kris.

Kris says that a big part of the reason bulk production is up at the same time efficiency is going down is because our "modernized" plants are getting worse at working with microorganisms in the soil. "So even though we've had all these great breeding programs and GMO technology we're actually getting less efficient crops," she explains. "And part of the reason we're getting less efficient crops is because we're selecting varieties through breeding programs or through genetic modification that are less efficient at working with microorganisms. We've overemphasized pushing plants to produce the highest yields, and so they're putting so much carbon energy into the aboveground portion of the plant, into making grain, that they have less carbon inputs that can go down through the roots and feed the biology in the soil."

Kris says the problem of ignoring soil biology is bolstered by industries, which profit from that ignorance. I ask her to give me an example of how this works. She answers from her experience working on her parents' farm.

"So as a conventional farmer, you go to your chemical dealer and he pulls out his little cookbook, which research and science helped come up with," she begins. "The chemistry in this little cookbook says for every bushel of corn you produce you have to put on a pound of nitrogen. So if you want two hundred bushels of corn per acre, you have to put on two hundred pounds of nitrogen. But at the end of the growing season you don't get two hundred bushels of grain. So you go back to your chemical dealer and he says, 'Well . . . you know. . . it was a wet spring, it was too dry in the summer, it was too cold, it was too hot, blah, blah, blah,' all of the regular excuses.

"You know that the book says for every bushel put on a pound, but you *could* just add more. Nitrogen's cheap so you add more. And this is where our efficiencies go down. We continue to add more and more and more. But the more we add, the more it puts stress on the relationships with the microorganisms." Herein lies the rub of treating agriculture like a 1950s chemistry set—American farmers have fallen into the "more is better" trap.

This strikes a personal chord with Kris, who emphasizes that conventional agriculture is "not gonna help you stay on that land. It's not gonna help you keep that farm. We are playing a chess game and chemical conventional agriculture is constantly allowing nature to checkmate us because we can't do better than nature."

Kris explains that prior to the past seven years, her father did not enjoy farming with the "chemical cookbooks" that agriculture colleges and extension agents push. She calls farmers "Renaissance people" due to their mix of knowledge and skills ranging from plant biology and pathology to mechanics and hydrology. Being involved in organic agriculture, she says she gets to see farmers come back to life. "They get a sparkle in their eyes, it's very exciting," she says.

COULD SOIL SAVE US?

Reducing the living world to numbers and chemicals has allowed humankind to build and do a great many things. But it has also hidden from us the effects of our industrial conquests. By looking only through the lens of inorganic science, the material, the man-made, we lose sight of that which flows, shimmers, runs, slithers, flutters, and beats. Reducing the world to which we are intimately tied to simple mechanics gives us the ability to control, manipulate, and dominate. But it robs us of connection. When one is connected to life, one is aware of its birth, its vivaciousness, and its death.

For the most part, the way we practice agriculture today is an extension of a mechanical view of the world. It is that same view that has shot our species forward on a path that has so far pushed an estimated 1,510 gigatons of carbon dioxide and correspondingly large amounts of oxides of nitrogen and water vapor into our atmosphere.[3] But using the thinking that brought us the Industrial Revolution, these numbers are just more chemistry to be "fixed" with yet another new technology.

However, if humankind wishes to deal with its atmospheric carbon dioxide dilemma, we are going to have to move beyond inorganic thinking and into organic thinking.

I ask Kris what she thinks about the idea of regenerative agriculture. She launches into her final lesson for the day. "There are two scenarios in which the planet can respond to storing the carbon that we currently have in the atmosphere," she begins. "The one which is easier for nature to do by itself is to put it in the oceans. The problem with putting that carbon in the oceans is that acidifies them. And for life on Planet Earth to be the way that we would want we don't want the oceans to get acidified." I am hoping that whatever she is going to reveal behind door number two is a more pleasant future.

Kris gets right to the point: "We have to put it back into the soil. And we have tremendous potential to be able to store that carbon in the soil in a relatively short period of time. The soil environment can hold more carbon than both the atmosphere and the organisms and plants living on the surface of the soil combined."

That's good news for the world's oceans, which have become 30 percent more acidic than they were at the start of the Industrial Revolution. Unless something drastically changes, by 2100 the global ocean acidity level is expected to more than double, threatening much of the life at the foundation of the planet's food pyramid and especially threatening the phytoplankton, which make more than half the oxygen we breathe.

Kris says that the Rodale Institute originally called organic farming "regenerative agriculture"; in other words, a practice of agriculture that improves soil, plant, water, human, and planetary health. "What we are trying to do in regenerative agriculture is create soil from the top down. We are trying to regenerate what was there and put that back in there," explains Kris. She says that since the initial studies in the United States in the 1990s, there's ever more evidence that organic farming outperforms conventional farming on nutrient-use efficiencies, stress management, energy use, and nutrient density. But the thing that may matter most in the long term is carbon sequestration.

Kris says that in their field tests they are seeing that carbon isn't just being brought down into the soil organic matter (SOM) in the first few inches of soil, but rather into soil fractions that will keep the carbon bound for a very long period of time. "If you want to increase the organic matter in your soil, pile a whole bunch of compost on top. I can guarantee you that will give you high carbon content in your soil, but that doesn't *sequester* the carbon," she explains.

"We want to create these systems in which the biology and the chemistry and the physics are all working together to store that carbon in that soil for decades to centuries, and that's what we're finding in the organic system compared to conventional." Kris believes that the way organic farming is being practiced today can be improved to sequester more carbon. She's also quick to point out that organic agriculture forms the foundation for what she believes will be the future of agriculture.

"We can still feed the world, we can still have all our toys," she says.

"We just have to change our minds. We have to stop thinking that the only solution is technology. We need to work with nature."

THE PEOPLE VS. THE INDUSTRY

On February 12, 2009, Tom Vilsack, then US secretary of agriculture, celebrated the two hundredth anniversary of the birth of Abraham Lincoln by establishing a USDA Certified Organic garden in front of the Department of Agriculture building in Washington, DC. At the time, the USDA Organic label had only been around for seven years.

According to the press release, to create the garden, 1,250 square feet of "unnecessary paved surface at the USDA headquarters" were removed. The "breaking pavement" ceremony, as it was cheekily called by the USDA, was symbolic of a move toward resource conservation.

"It is essential for the federal government to lead the way in enhancing and conserving our land and water resources," said Vilsack. "And one way of restoring the land to its natural condition is what we are doing here today." In honor of Abraham Lincoln founding the Department of Agriculture and calling it "the People's Department," Vilsack and the team at USDA created what they called "the People's Garden."

At the time, the USDA had planned to install a similar community garden at each of its ninety-three Foreign Agricultural Service offices in seventy-two countries worldwide.

To ensure their garden got off to the right start, they requested a truckload of Grade A, USDA Organic compost from none other than the Rodale Institute. Rodale happily donated the compost. As Jeff Moyer, the farm director at the institute, said, "Compost is a big part of what makes organic farming work so well."

Now sitting with Maria Rodale, I ask her how the garden is doing. She says she was at USDA headquarters the day "the People's Garden" was unveiled.

"That's where I met Secretary Vilsack," she recalls. "I had him sign my copy of the Farm Bill, and I said, 'My grandfather is shouting in victory from his grave that there's an organic farm in the front yard of the USDA in Washington on the Mall,' " she says with a twinkle in her eye. "But the last time I walked by it, all words about organic had been removed."

Maria says this type of backpedaling by the USDA is not uncommon because "there's tons of pressure from industry against organic." Sequestering carbon and growing healthy food without chemicals may be good for the people and the planet, she says, but it's bad for the bottom line of the businesses that manufacture toxic chemicals.

"If I was to say anything to the people working in the GMO and chemical industries, it would be: Look around you. Go out in the woods and appreciate nature. Listen to nature. And do the right thing. Do the right thing for the future. Do the right thing for your kids. Do the right thing for the world. Chemicals include the fungicides and herbicides and insecticides and fertilizers that, when combined, destroy the life in the soil. And the life in the soil is what holds the carbon. And it's what creates healthy, long-term food for all of us. *So don't f*** it up!*"

Maria has a good laugh about getting too carried away. But she says I can still use her quote.

If anyone has a right to be protective of the soil, it's Maria. She's carrying the burden of three generations of people who, with the evidence that organic food is critical for humanity's survival, have promoted it, fought for it, died for it, and continue to work for it. In their own way, the Rodales have carried the torch of telling a much older story, one that reaches back to the birth of human civilization. It's a story Maria's grandfather J. I. Rodale knew well. He saw how healing the soil could provide humanity with healthy food, a healthy ecosystem, and the basis for long-term civilization. He experienced firsthand the power of life to create life.

Perhaps we should all get a little more carried away about keeping the chemicals off our land and out of our food.

After all, it's only life as we know it that's at stake.

CHAPTER 5

THE BUFFALO BANK ACCOUNT

I am driving westward from Sioux Falls, South Dakota, into a vast expanse of rangeland. Once off the interstate, the world I encounter is largely devoid of structures, humans, and beasts. My vehicle is moving forward but it appears as if I am standing still. I have become a tiny speck traveling inside an ocean of grain. Rows and rows of a monocrops stretch to the horizon in all directions, giving the illusion the world outside is in a suspended animation.

South Dakota is in the northern "Spring Wheat Belt" and produces astonishing quantities of commodity grain. In this state alone there are almost 5 million acres of corn, around 4 million acres of soy, and 3.5 million acres of hay. Almost all these crops, along with the neonicotinoid insecticides and the dioxin-based herbicides sprayed on them, will be fed to animals, which in turn will be fed to us. Like a good soldier in the battle to feed the world, this state dutifully plays its part in the Corn-CAFO complex.

For the carbohydrate-based foods found in bags and boxes in our supermarkets, this state also grows almost 3 million acres of wheat. Corn, soy, hay, animals, wheat—these are the toes on the ecological footprint of postmodern man. Like the civilizations that preceded ours, our cultural history is being etched into this land by our food system. Because our culture places such little significance on ecological history, it is easy to forget that this "monocrop world" is a recent creation.

A 1940 Soil Conservation Service map of this area shows a radically different ecology from the one I now observe. The man-made Midwest of single-species agriculture was heretofore a prairie thick with perennial grasses whose roots often stretched ten to twenty feet deep into the soil.[1] The eastern frontier of this state was covered in Kentucky bluegrass, western wheatgrass, blue grama, dry-land sedges, sideoats grama, little bluestem, and sand dropseed. Toward the Black Hills in the west, the mixture gave way to grasses with names like "needle-and-thread" and "buffalo grass."

These prairie gasses were the "carbon pump" that drew carbon dioxide from the atmosphere and sent it deep into the soils, where microbes fixed it into stable forms. That dark, rich soil is the basis for America's great "breadbasket." As in all ecosystems, the grasses were part of a larger symbiotic relationship. Prior to European settlement, there was also another organism on these plains, one that both depended on the grasses and without which the grasses themselves would wither. The symbiosis between this creature and the prairie grasses gave us the fertility from which modern agriculture now draws. Designated by its scientific name, *Bison bison*, the American bison is better known as the buffalo.

The American version of this beast is distinctly smaller, wilder, and more forage-oriented than those bison that evolved in Europe and Asia. Aside from their fur and meat, these majestic creatures have some unusual attributes. They can leap vertically as high as six feet into the air and can run as fast as forty miles per hour. Consider for a moment the physics of a two-thousand-pound animal running at the speed of an automobile or levitating at the drop of a hat to the height of a grown man and you may begin to appreciate the exceptionality of these animals.

The muddled attempt at a European name for the bison would have been easier had the settlers just asked the locals what it was called. Their answer would have been "tatanka." In terms of the tatanka on the plains when the white man arrived, estimates put their numbers between 25 million on the low side and 60 million on the high side.[2] Given the vast area they traversed, their density, and the numbers of carcasses counted during their massacres, it is likely they numbered at the higher end of these estimates.

As herd creatures, bison would range the land in large packs of many thousands. They would move north in the summer and south in the winter, followed by, and constantly culled by, predators. Their area of habitat on the North American continent was vast and stretched from northern Florida to Vermont, to as far northwest as Alberta and as far west as Northern California, then south down through Texas to the Sierra Madre Occidental mountains in Mexico, which formed the southern border of their yearly journey.

It is said that when a bison herd moved across the plains, the thunder of their hooves could be heard from miles away. As they passed, the herd formed an impenetrable wall of animals lasting for up to two days. In 1842 Philip St. George Cooke was traversing the Santa Fe Trail when he wrote the following account:

> Suddenly a cloud of dust rose over its [the hill's] crest, and I heard a rushing noise as of a mighty whirlwind . . . I had not time to divine its cause, when a herd of buffalo arose over the summit, and a dense mass, thousand upon thousand, galloped, with headlong speed, directly upon the spot where I stood. Still onward they came—Heaven protect me! It was a fearful sight.[3]

In 1859, Horace Greeley was traveling from New York City to San Francisco when he had a similar experience. He wrote:

> What strikes the stranger with most amazement is their immense numbers. I know a million is a great many, but I am confident we saw that number yesterday. Certainly, all we saw could not have stood on ten square miles of ground. Often, the country for miles on either hand seemed quite black with them.[4]

To put a fine point on it, 10 square miles is 6,400 acres of land, a heckofa lotta space. As the largest group of wild herd animals to have

ORIGINAL MIGRATION TERRITORY OF PLAINS BISON

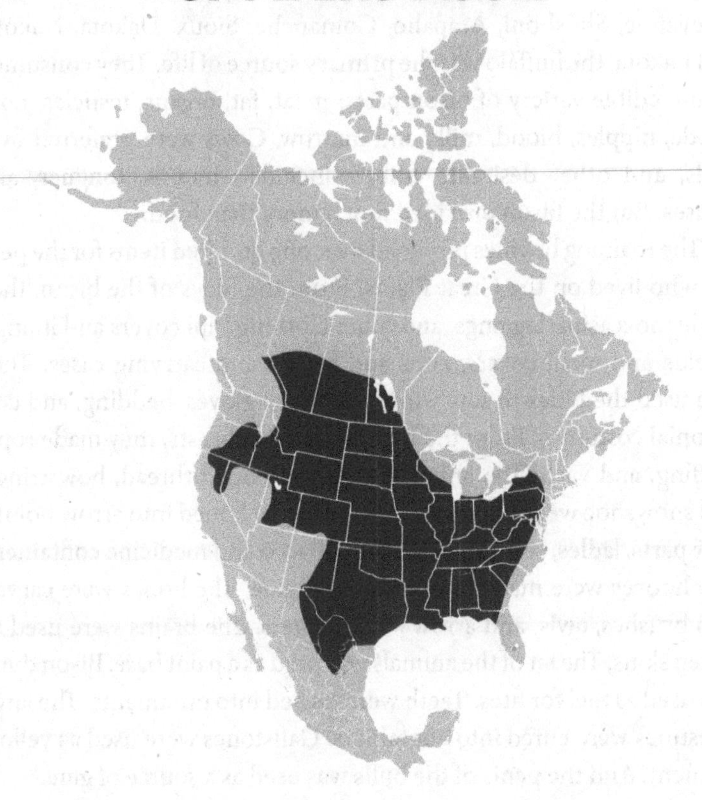

Not that long ago, bison (aka "buffalo") maintained the healthy soils of America through their constant migration.

existed on Earth since the retreat of ice some twelve to fifteen thousand years ago, the firsthand experience of a moving heard of buffalo must have been awesome. These beasts and the fertility they bestowed on the land created the soil bank account from which almost every American still eats, or rather withdraws, to this day.

TATANKA TRIBES

For numerous Native Americans, including the Blackfeet, Crow, Cheyenne, Shoshoni, Arapaho, Comanche, Sioux, Dakota, Nakota, and Lakota, the buffalo was the primary source of life. They consumed an incredible variety of bison parts: meat, fat, organs, testicles, nose gristle, nipples, blood, milk, and marrow. Cows were preferred over bulls, and other desirable edibles included humps, tongues, and fetuses. But the bison provided much more than food.

The roaming bovines provided over one hundred items for the people who lived on the Great Plains. From the hides of the bison, they made moccasins, leggings, and other clothing, tipi covers and linings, shields and maul covers, cups and kettles and carrying cases. They also used the hides to sew winter clothing, gloves, bedding, and ceremonial costumes. From the hair of the great beasts, they made rope, stuffing, and yarn. The animals' sinew provided thread, bowstrings, and snowshoe webbing. The bull horns were honed into arrow points, bow parts, ladles, spoons, cups, and tobacco and medicine containers. The hooves were made into rattles and glue. The bones were carved into brushes, awls, and arrow straighteners. The brains were used to soften skins. The fat of the animals was used as a paint base. Bison dung was used as fuel for fires. Teeth were turned into ornaments. The large intestines were cured into containers. Gallstones were used as yellow pigment. And the penis of the bulls was used as a source of glue.[5]

Prior to European settlement, the men, women, and children of the American continent lived their whole lives from birth to death in contact with some element of the American bison. And many of them would have been involved in the buffalo hunt.

Hunting buffalo was deeply embedded in the Plains Indian culture and was intertwined with their spirituality and rituals. Singing, dancing, sacred stones, and tobacco smoking were often part of ceremonies that preceded the hunt. Such actions were used to "call in" the spirit of the buffalo.

The act of hunting was often done by a lone man who was disguised inside a buffalo hide or wolf pelt. This "animal impersonator" would ease his way close enough to the bison to shoot one with a bow and arrow.

Another common hunting method was to chase a herd into a ravine, off a cliff, or into a snowbank. This "herd hunting" method often resulted in large numbers of buffalo being killed. While the Plains tribes generally used every part of the bison they killed, some early Europeans reported that the Native Americans were wasteful and left many unused carcasses to rot. This is likely a result of not being able to stop a herd of buffalo from running over a cliff once the "chase" had begun. The overkilling of a couple hundred animals was nothing, however, in comparison to what was soon to come.

OF MACHINES AND MEN

Just as the lives of the Native Americans and the buffalo were joined, so too was their shared demise inextricably linked. The Louisiana Purchase in 1803 and news from the Lewis and Clark expedition from 1804 to 1806 fueled waves of European immigration to the middle of America. By the time wagons rolled into the Great Plains, between one-half and two-thirds of the Native American population in the United States had already been wiped out by European viruses and diseases for which they did not have antibodies.[6] Skirmishes and battles were common between the settlers and indigenous peoples.

Military forts were soon established across the Great Plains, giving sanctuary to a new type of person—the commercial buffalo hunter. Because the American military "bluecoats" were often at war with the Native American tribes, buffalo hunting became a way to debilitate the "redskins" by destroying their primary food source. Just as the buffalo had provided the Native American tribes with needed products, they now provided the US Army with meat, machinery belts, and army boots. With firearms and horses, the slaughter of hundreds and sometimes thousands of buffalo at a time became routine.

The federal government promoted bison hunting as a means to create rangeland for cattle, to weaken the Native Americans, and to pressure them to move onto the reservations. By the 1830s the military, fur traders, and commercial buffalo hunters had enticed certain tribes into the business of mass slaughter. It is thought that by that time, the Comanche and their allies on the southern plains were killing about 280,000 bison a year, which was pushing the limit of sustainability for that region. A growing export market for buffalo robes and bison meat resulted in larger and larger numbers of bison killed each year.

By 1840 the number of buffalo had dropped from an estimated 60 million to about 35 million.[7, 8] But the biggest culling was still to come. Drought soon hit the Great Plains, increasing pressure for precious food and water resources. Then came the transcontinental railroad. Under the Pacific Railway Act of 1862, railway companies were given money and land to lay the tracks that would eventually connect the East and West Coasts. Along with the workers and the trains came a boom in buffalo hunters.

As railways expanded they hired mobs of commercial hunters. Among those hunters was William Frederick "Buffalo Bill" Cody, who was hired by the Kansas Pacific Railroad to supply the laborers with a steady supply of buffalo meat. Cody used a large .50-caliber Springfield Model 1866 rifle that he called "Lucretia Borgia" to kill his prey. One account said that he killed 4,282 buffalo in an eighteen-month period in 1867 and 1868. With word spreading of Buffalo Bill's legendary killing spree, money to be made, and mouths to feed, hunters arrived in throngs.

The trains themselves became effective hunting platforms. When a train approached a buffalo herd, it would slow down to match the speed of the herd. Using guns kept aboard for defense (or offense?) against Native Americans, men would shoot the buffalo from atop the train and out of the windows. Railroads were in favor of culling the buffalo, which could damage locomotives and halt trains for days on end.

As mountains of buffalo carcasses piled up, a movement began to protect the animals. Among those who spoke out in favor of protection was none other than Buffalo Bill. Despite pressure to save the species, in 1874, President Ulysses S. Grant vetoed a federal bill designed to protect the buffalo. Many powerful leaders in the government and military were firm in their stance; the buffalo should be extinguished and with them the American Indians' source of independence. As the culling continued, poorer famers harvested the bison bones and sold them for fertilizer. By 1900 the once great species that had dominated the American heartland for thousands of years had been systematically reduced to around three hundred animals.[9]

Indigenous tribes were relegated largely to barren lands far away from any remaining buffalo. With their food source and central cultural icon all but extinct, Plains Indians were forced to attempt to farm and rely on government rations for food. Lieutenant Colonel Dodge is reported to have said: "If we kill the buffalo, we conquer the Indian." Indeed, in its rush to fulfill its "manifest destiny" and expand westward, the United States had done just that.

The destruction of the buffalo and elimination of their roaming and grazing patterns radically altered the ecology of the Great Plains. As the native grasses began to oxidize and die, topsoil erosion increased. This helped set the stage for the Dust Bowl of the 1930s.

Today, agricultural and residential development of the prairie is estimated to have reduced the prairie to one one-hundredth of its former area.

PINE RIDGE

By the time the Civil War was over, the US government had enacted four hundred treaties with the Native Americans, most of which it would later break. Among the broken treaties are the 1851 and 1868 Fort Laramie Treaties, which in effect granted the Sioux and Lakota nations a portion of land equal to the size of North and South Dakota combined.

Their legal territory was centered on the Black Hills and encompassed much of what is today South Dakota, about a third of Nebraska, almost half of Wyoming, and chunks of Montana, Colorado, and North Dakota. In a landmark 1980 federal court case, *United*

States v. Sioux Nation of Indians, the Supreme Court upheld the validity of the treaties and offered the tribes $120 million for title and back interest to their land. The tribes refused payment, demanding instead the return of their native lands. The case remains open.

In the Lakota language, *Wazí Ahánhan Oyánke* refers to the Pine Ridge Indian Reservation. Located in the Badlands of southern South Dakota, this area of land is a tiny fraction of what, according to the US Supreme Court, actually belongs to the Lakota and Sioux peoples. Today it is all that remains of the once thriving nation of millions that encompassed the Lakota, Dakota, Nakota, and Sioux nations. About thirty-eight thousand Lakota people live within the borders of Pine Ridge.

Being invited onto "the rez" is an act of trust. The Americanized name of the man I'm meeting is Tony Ten Fingers. He's the author of a book called *Lakota Wisdom*. The book contains a series of quotes superimposed, Instagram-style, over beautiful images of the world of the Lakota. It is simple and moving.

Pine Ridge is a rough place. The site of two separate gun battles in the 1970s between activists from the American Indian Movement (AIM) and the FBI, the main settlement still looks like a war zone. There are no street signs. The roads are no more than dust pathways. Derelict cars and house trailers sit haphazardly across the township as if dropped at random by a tornado.

After numerous failed attempts, I locate Tony's house. He's a tall and broad man whose face is scarred by the ravages of a difficult youth. His black hair is pulled back in a ponytail. We sit for a while outside and chat. He asks if I wouldn't like to do our interview somewhere with a deeper personal meaning to his people.

We head up the road for about twenty minutes and pull off at a lookout. I follow Tony toward a cliff edge and have to pinch myself to make sure I'm not dreaming. What stretches out before us must be a hundred miles of deep ravines carved into the Earth's crust. The expanse and beauty of this sight easily rival those of the Grand Canyon. The rocks are pink and purple and bluish. It is breathtaking.

"We are in a place that's very sacred to my people because it's a place of feeling safe. Of being comforted once more. This is what we call the Stronghold Table area," says Tony. He tells me that these ravines have been a hiding place in times of danger for many Native American leaders, some of whom went into the ravines never to be seen again.

Tony tells me that his real name in Lakota is Charging Eagle of the Oglala Lakota of the Seven Council Fires, but that name is not used much outside tribal circles. When I ask him about his people's relationship to the buffalo, he thinks for a moment and unravels a history not found in textbooks.

"The buffalo meant everything to us because it literally nurtured us. Cared for us. It became a part of our ceremonies. There is this story of the creator, Tunkasila, telling our people to follow the buffalo and the buffalo will show you where your lands are. The Black Hills is this central area and when we migrated with the buffalo it was a huge circle that went up into Canada and all the way down into Colorado Springs, all the way out to the Tetons, the Rockies up north, and back into Canada. This whole area, this whole region, belonged to our people—the Lakota, Dakota, the Nakota nations."

It had never occurred to me that buffalo might have actually migrated in a circular pattern. For example, the herds followed by the Lakota moved around the Black Hills as their center. Given the changing of the seasons, the need to allow the land to rest for a year, and the need for the herds to keep moving, this grazing pattern makes sense.

"When you use an animal to nurture the people, you become that animal. You become what you eat," says Tony. "Even today, when you look at some of the foods we eat, we can almost tell who is what. So if you eat a Big Mac, you probably look like a Big Mac," he says, laughing.

"But we followed the buffalo. We ate the buffalo. We became the buffalo. The buffalo was honored and used in our ceremonies. We had buffalo dances. The buffalos constructed our homes—the tipi. We slept in the buffalo robe. In the winter it kept us from the bitter chill and kept us warm. And they took care of us. So the buffalo was our world and we were known as the buffalo people," he says.

I ask him how he thinks his people differed from people of today in their view of the natural world. "We didn't see ourselves as above animals or below animals, we were just there with them. We became a part of that bigger system of life—the circle of life. When I talk about my ancestors they lived in a period when you can actually drink from the rivers. You can actually eat what was caught in the rivers. You can actually gather berries. And everything came from nature. Everything was a part of this bigger circle."

Tony says this ancestral way of being stands in direct contrast to what he calls our "accumulation culture." "When you're born and raised in what I call the accumulation culture, you lose sight of the bigger picture, because when you start to accumulate, you tend to forget what is important, and what matters shifts, and you become separate. And so the more you become separate, the unhappier you're going to become. Accumulation is like an addiction. There is this need to want more, and more, and more. There is no satisfaction. You want to earn. You want more material things, and once you have more material things, you're going to want more. There is no end to that."

Tony tells me that our wars, our destruction of natural resources, and our degradation of the land and climate all come down to our misplaced priority on the accumulation of material things and the resulting separation we have from the natural world. "It's all part of this illness, this addiction, that prevents us from being sustainable," he says.

In terms of the destruction of his people by Europeans and Westerners, Tony is careful not to place blame. "What happens when you don't learn from your past, you're not going to progress. You're going to learn newer ways of doing things, faster ways of doing things, but yet you're not learning what really matters. That's why things are being repeated over and over. History repeats itself because we haven't learned anything."

Tony tells me that his interest in learning and teaching about his culture's connection to the Earth was a result of a violent car crash he was in. His heart stopped beating and he was close to death. "I didn't see a bright light or anything like that. But I did see children playing," he says. It was then that he realized what is at stake for all peoples—our future. He says he now spends his time working with disenfranchised youth across the country in order to connect them back to nature. "I talk about our relationship with the Earth and with the world and with one another," he says.

"And so I think with being in an accumulation culture, we have to make that shift once more back to our connection and I think that's where we're at right now. And it can be done. I mean . . . I believe it can be done."

On the drive back to his house, I ask Tony, "What happens if it can't be done? What happens if we don't learn from our mistakes of the past and if we cannot reconnect to our natural place in the world?" He tells me that many of his people believe this may come to pass. They are prepared for society as we know it to collapse. For them, it will simply be time to go back to the "old ways" of hunting and living off the land.

By the time we make it back to Tony's place it's getting dark. I bid him farewell and head to the closest hotel, the Prairie Wind Casino & Hotel. Past the smoky and empty slot machines I find a restaurant. As I eat, I notice the scaffolding on one side of the room is holding some cans of paint. The mural that is in progress and will soon cover the upper walls of the restaurant is a scene of this area pre-European settlement. There are indigenous people and tipis and hunting, but the most prominent feature, painted with care, is the buffalo.

It is truly a moment of standing inside the "accumulation culture" looking back through a window at the past. It took "new Americans" only a hundred years to profoundly decimate the ten-thousand-year reign of the 50 million or so indigenous people and 60 million or so buffalo in the Americas. Do cultures with "superior" war technologies often decimate ones with "inferior" technology? Yes. Should we feel guilty? Perhaps.

But the bigger lesson is that war technology is not a foundation for ecological understanding. Nor does its presence secure a civilization's future. Our weapons have radically advanced since the European settlers drove into the prairies. What remains to be seen is if our ecological understanding can also advance.

America has fertile soil because for over ten thousand years buffalo herded over the middle of the country. Vegetarian or omnivore, when we eat, each of us withdraws fertility from that "buffalo bank account."

But now that account is running desperately low and we must do whatever we can to restock it.

SAINT FRANCIS OF ASSISI

Any attempt at rebuilding the buffalo bank account will likely come from within our "accumulation culture." We designed our cities and built our modern edifices around chemistry and physics. In that process, we ignored the biology that gives us life. It is now time to flip our worldview on its head. Now we must design from a biological perspective. The problem is, we still live inside the structures of mechanical thinking. Bringing our hard world together with the dynamic needs of ecology is challenging.

The city of San Francisco is named after the early thirteenth-century friar Saint Francis of Assisi. Saint Francis was a naturalist who is said to have communed deeply with plants, trees, and animals. As such he is revered as a protector of the natural world and is often pictured with a bird alighted in his open palm. Today, the city that is his namesake is also trying to do a little something to protect nature.

It's 5 AM. As I pull up to a Coffee Bean & Tea Leaf in the hilly Richmond District, the city is dead calm. I am here to meet a man who's going to teach me how San Francisco is trying to emulate nature by turning its food scraps into soil. In nature, you see, soil depends on decomposition of organic material. But in cities we throw our food scraps away. As the city that brought us free love in the 1960s and more recently the LGBT movement, San Francisco is now leading the world with another alternative lifestyle choice.

I'm talking about compost.

I immediately spot Robert Reed sitting at a table wearing a crisp white button-down shirt and a yellow safety vest. Sitting with him is another man with a safety vest and a woman with a video camera. I say hello, and Robert, who is delivering the most intensely animated 5 AM monologue I've seen since college, tells me to wait a few minutes as he's explaining something "important." I get a chai and sit at a nearby table. Robert is speaking so loudly that the entire coffee shop reverberates with his explanation of how the garbage trucks pick up food scraps for compost. I consider getting back in the car and going back to the hotel, but it's too late now.

Robert calls me over, and I am introduced to everyone. There's Tom, his director of operations; Anne, a French journalist also covering this story today; and Robert himself, who is the public relations manager for a company called Recology, which is in charge of picking up and disposing of all the waste in San Francisco.

Robert, who stands about six-foot-four, with a head so shiny it might squeak if you rubbed your palm on it (I didn't try) and a good-sized gap between his front teeth, seems to have exactly one speed—fast-forward—and exactly one volume—max. He explains that this morning is all about safety, because people are accidentally killed by garbage trucks with surprising regularity. (Don't believe him? Google it.) He explains (loudly) that one misstep and the sixty-four-thousand-pound truck with a thousand pounds of per-cubic-yard hydraulic power will "crush you like a grape." Point taken.

Robert hands me a yellow safety vest, goggles, and a brightly painted helmet and gives Anne and me the garbage-truck-safety-101 briefing. Then he looks at his watch and says, "We gotta go," and without another word darts out the door. We all follow him into the wet San Francisco morning air. Robert stares intently at his watch and shouts, "Your truck should be coming down this street . . . right . . . now." Just then an unmistakable thundering sound approaches as a garbage truck rounds the corner. One thing's for sure about Robert, the man knows his business.

I am introduced to the driver, Jesse Chavez, and Robert explains that Jesse isn't there to babysit me and tells me again how to not die by garbage truck. With that we're off, all the way down the street to a house twenty yards away.

Jesse stops the truck, pulls out a massive wad of keys, jumps out, walks over to a door next to a garage, unlocks it, and comes out rolling a green trash bin. He maneuvers it over to the side of the truck, uses some hydraulic levers to control the "arm," which grabs the bin, lifts it up, and dumps it into a special section of the truck just for compost waste. He then takes the bin back, moves the truck ten feet, and repeats the whole process. This is Jesse's morning. The houses change, but his actions are the same.

Jesse and his truck are the first step in a process that results in mountains of compost for farms and vineyards around San Francisco. That compost is critical for growing food on organic farms. That food, in turn, finds its way back onto plates in restaurants and dining rooms around San Francisco.

This urban composting program is by no means a perfect closed-loop cycle. But the fundamental idea of turning various forms of biodegradable waste into the basis for our food chain could be one of the most crucial components in rebuilding a safe, effective, and sane food supply for our planet. And as J. I. Rodale pointed out, compost is a good place to start.

WASTE NOT, WANT NOT

Picking up green bins from each of San Francisco's over 870,000 residents was not an overnight success. While Recology has been making compost in the city since 1995, the idea to make it mandatory came from a 2005 initiative spurred by then San Francisco mayor Gavin Newsom. Newsom, who as of 2017 serves as California's lieutenant governor and who always manages to look like he just stepped off a movie set, has a penchant for environmentally oriented work, especially when it comes to the intersection of food and climate change. He saw that making compost from food scraps could have a positive impact on both fronts.

In 2005, Newsom invited mayors from around the world to San Francisco, where he ceremoniously announced that San Francisco would be zero waste by 2020. With typical Gotham City flare, Newsom was setting the high bar for a program called Waste Zero. Not only was he committing San Francisco to zero waste, he was enticing other mayors to do the same.

Waste Zero sounded like a good way to reduce emissions, put carbon and water back into the soil through compost, and put San Francisco on the map as a leader in sustainability. But executing the idea was far more difficult than putting on a conference of mayors. According to Newsom, "Almost seventy percent of San Franciscans live in apartment buildings. If you think what we did with same-sex couples in 2004 was controversial, when we required composting in San Francisco people went completely insane." Newsom is one of those rare politicians who can get away with making a public statement that mixes homosexuality with compost.

The lieutenant governor says that "what makes San Francisco's composting and recycling program work is we incentivize people to keep stuff out of the black bin. When you have nothing in the black bin we don't charge you. When you have a ton of stuff in the black bin we charge you a lot. So your goal is to get it into the green and blue bins—into the composting and recycling bins." He adds, "You wanna move the mouse? You gotta move the cheese."

The combination of a disincentive for landfill waste and personal green bin concierge eventually won San Franciscans over. Today San Francisco's compost program collects waste from twenty-five massive hotels, five thousand restaurants, and about 350,000 homes. The green bins are filled with anything that came directly from nature, including food scraps, coffee grounds, yard clippings, and biodegradable paper products. All told, San Francisco produces about 650 tons of compostable material per day.

Newsom's vision is becoming a reality in other ways as well. Recology has a massive recycling program that deals with everything from building waste to hard goods to paint to batteries to the usual aluminum, glass, and plastic containers. Each year, San Francisco composts more, recycles more, and throws less into its landfill.

MORDOR IN THE MORNING

Driving to Vacaville, California, in July is an exercise in inhospitable thermal dynamics. The small, dusty city is located in a kind of no-man's-land between San Francisco and Sacramento. And in the summer, the heat is blinding. The temperature outside the car reaches 100 degrees Fahrenheit as I make my way out of the Bay Area. In the late afternoon when I finally locate my el-cheapo hotel in a sea of asphalt off Interstate 80 called Fairfield, it's 110 degrees and climbing.

Robert Reed has once again given me unpleasant instructions. I am to meet him in front of the Days Inn at 4:00 AM sharp. And he's not going to wait for me if I'm not there.

Like Jesse Chavez, my alarm goes off at three thirty in the morning. At four I am sitting in the freezing-cold wind in front of the hotel. I wait for forty-five minutes, contemplating going back to bed. When I can no longer feel my hands, Robert pulls up in his Prius with Anne, the French journalist. They both have large lattes. Urgh.

We head out of town through the wind and toward a facility that looks like something from the movie *Dune*. Bright lights pierce the darkness in all directions. Machines the size of four-story buildings lumber among all forms of refuse. When the wind occasionally switches direction, we are enveloped in a stench so wretched it makes me glad I haven't eaten yet.

Robert checks us in at a security office. I ask if there's perhaps a jacket I can borrow. He gives me a fluorescent parka with the name "JOE" stamped on it, and judging from the size of the thing, "Joe" must be a big dude. Anne meanwhile gets to be "Roger."

Robert takes us into the yard and into another world. Mountains of fetid garbage glow under halogen lights. Larger-than-life yellow Tonka toys scrape, push, dump, and pick up various piles of refuse. Men in fluorescent garb work like ghosts in the half-light.

Robert explains that this is the "sorting" area. A huge semitruck backs up onto a ramp and dumps its contents on the ground below. Men and machines begin to move the mess into piles. A front-end loader picks up a bucket of the stuff and dumps it into a huge revolving cylinder called a "sifter-sorter." From there, the stream of mostly biological waste moves onto a conveyor belt that takes it upward and through an open-air, second-story "room." It's here that men who are covered from head to toe in reflective gear are physically pulling plastic and other noncompostables out.

"Everyone should have a chance to work in this room!" screams Robert over the oppressive machine and wind noise. We climb the metal stairs, stand in the sorting room, and watch the men sorting. As I stare at the guys working under the intense lights pulling out the bits and pieces of garbage from the green bin waste produced by the good people of San Francisco I feel like a voyeur. Yes, these men have jobs in a company that is employee-owned and they are doing important work. But I've seen prisoners with cushier jobs. This is the bottom of the barrel, literally.

The reason Robert wants everyone to get a chance to work in this "room," he says, is because "sometimes people make mistakes" by putting the wrong thing into the green bin. Like parts of the machinery itself, the sorters pick out a steady stream of plastic bags, Styrofoam, and even condoms. It's clear that "sometimes" means that more often than not, people are careless with what they put into their green bins. Most folks probably have no idea that human beings have to pick through this stuff to remove their "mistakes."

We follow Robert back downstairs to a growing pile of post-sorted food scraps. Anne, meanwhile, is asking questions, managing her camera, and simultaneously smoking a cigarette. It's an impressive juggling act, even for somebody who isn't inside a garbage dump. Robert is very proud of the sorted pile of food scraps and explains that this is the "good stuff," the banana peels and things that will soon make the best possible compost.

The sun is just cresting the horizon and the wind bites harder into my exposed flesh. Robert takes us from the pile of scraps down an alleyway through mountains of black, smelly stuff. He explains that the whole process from bin to finished compost takes sixty days. As we walk he points out various piles in various stages of decomposition. The majority of work here, he explains, is not done by men or machines but by microbes. These microscopic agents work to break down the biological waste into something that is perfect for use on farm fields.

This place is as big as ten football fields. Pile after pile, mountain after mountain, the microbes and the waste churn and turn this stuff into farm-ready fertilizer. We hike past a machine that looks like a tractor that ate twenty other tractors. The mega-machine slowly passes over a long pile of compost, aerating it in the hazy morning light with its massive churning blades. I have to hold my jacket to my mouth and squint to avoid inhaling the stuff.

Eventually, we find our way to the end of the "line," where rich piles of sellable compost sit waiting for pickup. Robert crouches down at the pile and shoves his hands in. "It feels very warm because it's still biodegrading," he says. "Just sixty days ago this was food scraps, now look at it—it's beautiful, carbon-rich, brown and black. Absolutely amazing, isn't it?" This is a man who deeply loves compost.

We give our jackets back at the main office and Robert drives us up to a ridge that overlooks the facility. From high above I can see that the place is actually divided into two halves. The left half is compost. But the right half is a landfill. Robert asks if we want to go see the landfill side. Why not? I'm already covered in filth.

Standing in the middle of the hot, muddy, putrid garbage dump makes me want for the freezing wind and compost piles. As we pick our way up a hill of garbage, Robert tells me, "The unfortunate reality is most trash or garbage gets incinerated or sent to a landfill. We need to change the model. We need to first go to composting programs, and then also go to recycling programs, and move away from incineration, and move away from landfills. We need to model nature. And one of the greatest ways to do that is to have urban compost collection programs."

We trudge through the mud back to the car. Die-hard-French Anne has given up her composure. As the garbage mud squishes over her boots, she shouts, "*Merde, merde, merde*! This place is *merde*!" Oblivious to her complaints, Robert continues, "If you landfill, you have to buy real estate. You have to go through an expensive permitting process. You have to line that landfill. Then you fill it with trash, and then it's full. And you have to do it again. You have all those costs. Additionally, land filling and incineration are polluting. But by composting, not only do you avoid those costs, but composting actually helps farms and helps society save money because you can grow thirty percent more food by applying compost."

Recology collects about 650 tons of compostable materials every day in San Francisco. After sorting, biodegrading, and aerating, that compost turns into about 488 cubic yards of finished compost. A cubic yard of compost weighs around 1,300 pounds, so that works out to around 283 tons of finished compost coming out the other end each day. The difference between the weight of the original material and the finished material is due to evaporation of moisture and removal of "unwanted" materials. Since Recology sells its compost to farms, orchards, and vineyards for between $12 to $16 per cubic yard, making compost is also good business.

In many ways, compost is an ideal fertilizer. It's balanced for the soil, it is rich with beneficial microbes, and it helps soil hold up to twice as much moisture. When applied to soil, the microbes in compost go to work like probiotics, helping to establish that rich humus layer of carbon.

COMPOST TO CARBON CAPTURE

I head north out of the city and into the upscale world of Marin. I'm going to meet a man named John Wick (not "the" John Wick of the 2014 movie of the same name), but rather an upbeat former construction project manager who got interested in the role of carbon after struggling to control the weeds on his family's ranch. The project he cofounded is called Marin Carbon and it fosters a coalition of scientists, farmers, and researchers who are all looking into better ways to do biosequestration through regenerative agriculture.

I meet John for tea one misty morning on his ranch. John's bright eyes are framed by a pair of thin-rimmed glasses, and his thick flannel shirt and short-shaved hair all add to his Boy Scout–like charm. I sit at his kitchen table, and he gets out a white board and starts madly scribbling.

After an hour-long presentation about carbon cycling through our atmosphere over the past eighty thousand years, John says that before he got into all this, "I thought of soil as dirt, as an inert thing that I would push along with my bulldozer. Now I understand that soil is a community of life. It has completely changed how I look at the entire world. I see it now as a living skin."

John and his wife, Peggy, had numerous failed attempts at improving the land on their ranch. He sprayed chemicals, and the flora suffered. He kicked off the cows that were there, and the soils began to erode. He tried introducing other plant species and so on. But all his experiments ended with further deterioration of the land.

In 2008 John brought together a group of scientists to try something new. They mapped out suitable plots of land on his ranch for their experiment. The plots were segregated into control plots, which would remain untouched, and treatment plots. Onto the treatment plots they carefully spread a half inch of compost. And then they waited five months. What they found when they went back to the plots was encouraging.

John grins from ear to ear as he explains, "We measured a significant amount of water in the treatment plots compared to the control and we measured a ton of carbon per hectare compared to the control. The disturbing news was that the control plots lost carbon during the experiment." A little bit of compost, it appears, goes a long way toward water retention and carbon capture. What is important, according to Wick, is the long-term impact of spreading this compost.

Says Wick, "The next year, and the next year, and the next year, and the next year—the next five years an additional ton of carbon came into the system on its own. So, that single bump of compost knocked the system into a higher function and from then on it held more water, which promoted more plant growth, which removed more carbon from the atmosphere." Given the global implications of this discovery, one can see why John is excited. Not only do microbes rescue carbon from the atmosphere, they create the infrastructure in the soil for holding water.

This discovery has since been echoed by others studying the phenomenon of soil health. The connection between water retention and carbon retention is strong. Says Christine Jones, an Australian scientist working on soil carbon, "Generally speaking, water and carbon move together. Their cycles are connected." If soil loses water, it also loses carbon, and conversely, if soil retains water, it also retains carbon.

Using life to create life is Rudolf Steiner's foundation for organic agriculture. And it remains the basis for any honest modern organic farming or ranching operation. If compost could be the "life" that is needed to create more life, then voilà!

There's only one problem. There's not enough.

The issue with compost is that once farmers use it, they get hooked on its benefits, and, given a limited supply, demand turns into competition. In San Francisco's surrounding region the battle to get compost has gotten fierce with vineyards, farms, and orchards all vying for their truckloads of "black gold." Filmmaker Greg Roden, who covered the San Francisco compost beat for his remarkable PBS series *Food Forward,* calls the contentious situation "the compost wars."

America can produce a lot of compost, and every cubic yard of that compost would contribute to enriching soils while increasing plant growth. But if we apply the example of San Francisco's compost program to the total population of the United States, we come up short.

Collecting and composting all the food scraps and yard trimmings in the entire country every day of the year would produce enough compost to cover about 2 million acres with a quarter inch of compost. This is a significant area, but it is only about 0.5 percent of our agricultural soils.[10] To make a lasting impact on our soils we would need more compost—lots more compost.

Here's a quick back-of-the-napkin sketch of how much compost we can produce in the United States from additional "materials."

About 2.47 million people die in the United States each year. At an average weight of 166 pounds, that gives us about 205,000 tons of "compostable material." If we add to that the 7 million dry tons of biosolids from sewage treatment plants, we get about 7.2 million tons of additional "compostable material."[11] Composting the 7.2 million tons gives us about 9.5 million cubic yards of finished compost.[12] That's only enough to cover an additional 288,000 acres of land, or about one hundred modern corn farms.

Even with all our sewage and dead bodies, under the most aggressive strategy, that type of compost could only cover a small land area in and around cities. To be fair, this would still be a marked improvement for our urban soils and would drastically reduce the amount of waste cities produce. There is no doubt that kind of compost is an important tool in our soil repair kit. But by itself, city-derived compost cannot change the climate, fix the water cycle, or heal a broken global farming and food paradigm.

So what will?

Dr. Jeff Creque is one of the scientists working with Wick in the Marin Carbon Project. A thoughtful and concise man with angular, elven-like features, he has spent his entire thirty-five-year pre- and postdoctorate career studying how farming affects micro and macro ecosystems. Dr. Creque explains: "We could not, even if we wanted to, produce enough compost to spread across the world's rangelands. Secondly, it would be very impractical to try to do that." The question, as he sees it, is how are we going to replenish the nutrients of the world's agricultural soils?

Our planet has 5 billion hectares of arable land. One and a half billion hectares is under cultivation for farms while 3.5 billion is grass, prairie, and rangeland. Of that total acreage, the majority is desertifying due to poor agricultural practices. Our ability to grow food is shrinking while our population is rapidly increasing.

To grow food in the quantities needed to sustain humanity, soil must be nourished. There are exactly two ways to do it: by biological means such as compost (or something like compost) or by chemical inputs. City-based composting is helpful, but cannot offer more than a tiny fraction of land coverage. Conversely, the agricultural-chemical experiment results in ruined land, poisoned waters, and an array of debilitating health conditions. That leaves us with just one way to put nutrients and microbes back into the soil.

As Dr. Creque puts it, "The obvious answer is the intentional and intelligent management of our livestock." In other words, we need to use the greatest compost makers alive today. I'm talking about cows.

JOHN OF COWS

Dr. Jeff Creque tells me that *biocybernetics* is a new way of looking at biology where all things are connected. For Creque, a farm is not a farm. Instead it's a "system" with many interrelating pieces. This is also true with a rangeland "ecosystem," in which numerous biological processes are constantly interacting. There's a "flow" of chemical compounds through millions, billions, or even trillions of biological organisms.

This dance of life in service of agriculture was first codified by Rudolf Steiner; then by farmers in Brittany; then by a French biochemist and farmer named André Voisin, who published three books in the 1950s and '60s; then by the oft-loved (and sometimes hated) Allan Savory; recently by "radical" farmers like Joel Salatin; and now by people like those working in the Marin Carbon Project.

Says Creque, "Our agriculture for many years has been all about manipulating the components of systems and we focus on individual nutrients or the manipulation of individual variables. But from a systems perspective that's an entirely erroneous way to look at the world. If we begin to look at our farming systems as systems, if we begin to look at our cultural ecology as a system, then a lot of our problems begin to make much more sense."

Creque contrasts this with our modern industrial agricultural system, which relies on chemical inputs. "What we're doing today is undermining the very ecology we're dependent on," he says. "The long-term prognosis for our survival on this planet given business as usual is very, very poor. The dead zones around our coastlines globally resulting from the discharge of sediments and chemical fertilizers and pesticides and herbicides from our current farming systems cover a huge percentage. We are literally killing ourselves with the food production system we have today. To suggest that we need that system to feed ourselves, I would have to describe as the ravings of a lunatic."

It doesn't matter what type of agriculture you do, Dr. Creque says, you will always alter nature. "The only way we can manage an ecosystem is by disturbing it," he explains. "Even allowing an ecosystem to rest from the historical disturbance constitutes disturbance relative to the previous pattern of management. So the question becomes: What type of disturbance, what frequency of disturbance, what intensity of disturbance, is needed to achieve enhanced carbon capture on that landscape?"

Both John Wick and Jeff Creque agree that humans often misinterpret the visual signals of what kind of "disturbance" is going to move a "system" in the right direction. Wick gives an example from his own misinterpretation. He tells me, "When we started management of this ranch we were a little conflicted. We'd eat meat, but we did not like cattle. We believed the cattle were destructive to the environment and the very first thing we did was evict the young man who had grazing leases to this rangeland." But removing the cattle had the opposite effect from what he had hoped to achieve. "That actually caused a cascade of collapse to the ecosystem," says John. From there, the weed problem on his ranch worsened.

After his compost trials and studies, John decided to take another look at cow manure. The action of manure and the action of compost on the soil are similar. Thus, Wick wanted to see if he had in fact pushed his "system" in the wrong direction by removing the cows.

Wick tapped Dr. Gary Andersen, a microbial ecologist from UC Berkeley, to figure out the gut bacteria of the animals that once roamed his land—and thus the best animals to bring back onto it. Dr. Andersen found that elk, pronghorn antelope, and bears had all once roamed Wick's Marin land and were now "missing."

John tells me that "on the phylogenetic tree, the gut bacteria in the elk are side-by-side with the gut bacteria of the cows that I'm introducing. This suggests to me that the manures I'm depositing are actually hosting populations of bacteria that potentially would harmonize with the rhizosphere of the native plants."

"We can now think of cows as a proxy or a replacement for an elk. The pronghorn antelope that would have been in the system are very similar to a goat in terms of how they interact with the environment. Then in this area historically there would have been black bears also rooting around among the trees and the roots of the trees. Pigs do a very similar function in an ecosystem."

From Andersen's research Wick realized he could begin to mimic some of the things that natural grazers once did on the land. Says Wick, "Stepping into a system that's now had a couple hundred years of very strong European influence where the native species aren't readily available, I could assemble a team of animals to start doing ecosystem functions for me."

In spite of the other possible choices, John says he ultimately settled on cows because of their availability and practicality. "We were able to find a herd of local cattle that we could borrow to bring in as a management tool to promote native ecosystem function," he explains. "A lot of thought went into when in the year we brought them in, how many we brought, how densely we gathered them and pushed them through the system, how long they were there, and then how long they were not there."

The results were promising. The soil performed as it did with compost and began sequestering carbon. But it also did something surprising in regard to methane. "What we measured was that when cow manure landed on the soil there was no methane signature from that because there are methanotrophic bacteria in a healthy soil system that take apart that manure."

Dr. Creque is firmly of the opinion that managed intensive rotational grazing (MIRG) cows offers a practical, large-scale way to give agricultural soil the valuable nutrients and microbes it needs. Indeed, a typical thousand-pound steer poops about sixty pounds a day. Provided it is being grass-fed and moved often, its poop will feed the soil with 0.31 pounds of nitrogen and 0.11 pounds of phosphorus.[13] While they are mostly still confined today, the 80 million cattle in the

United States produce 4.8 billion pounds (2.4 million tons) of manure a day. Compared with the maximum potential of only 106,000 tons of city-derived compost, there's a lot more compost available from cow manure.

Working together, Dr. Creque and Wick have developed their own system of managed grazing. Explains John, "When the herd moves onto the field they all just start eating as a mob, and they walk across the field uniformly. They'll eat the desirable plants, the annual plants, and the exotic plants all at the same rate, and then we go to the next field. I'm averaging out the grazing impact over the field and we won't go back for an entire year. In three years we started seeing a shift in the

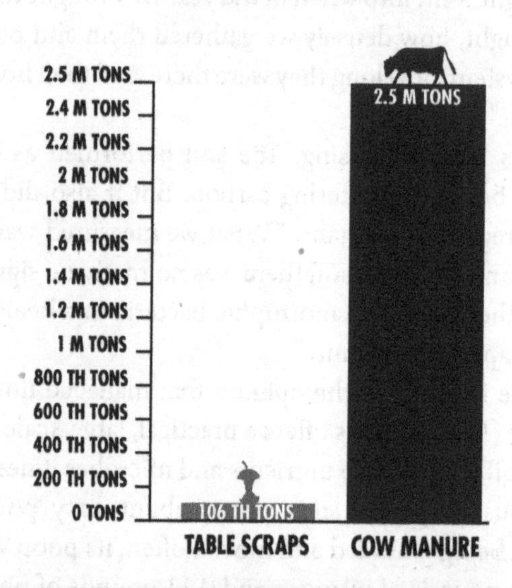

Ultimately, we need cows and the manure they produce to make enough compost to stabilize our agricultural soils.

grass composition toward the desirable plants. It was very exciting. Now after seven years of this kind of grazing management we've established a very significant native plant population."

There are two crucial differences between city compost and free-range cow manure that steered John toward using cows. The first difference is that while a national compost program is a good idea, the manure in question already exists. Certainly, the way it is being produced today is toxic (never mind the deplorable conditions of the animals). And certainly if it could be distributed naturally (by cows walking in pastures), it would be better than creating lagoons of waste (as is the case in most animal feedlots today).

But the second big difference is also critical. Manure does not need to be (and should not be) spread like butter on toast across the land. Instead, because of its potency, it is better for the soil if cow manure is deposited at relatively consistent intervals across the land.

John says his philosophy stems from the event of cow manure hitting the soil. He explains: "What I try to achieve is a cow pie every three square feet. I want to maintain that edge where I'm always restoring balance and restoring carbon to the soil rather than extracting it or depleting it."

John Wick does not own any cows, but rather "rents" them at his leisure to graze. His is a valuable experiment, but it is an experiment nonetheless. For mob grazing to work at scale, somebody has to own the cows and that somebody has to be responsible for the birth, life, and death of the cows.

Most of today's "mob grazing" farmers rely on feed grain and hay from outside the farm. By offloading the responsibility of growing the primary input for their livestock (and the cows' primary use of water), their "system" is incomplete.

But if cattle and other livestock are going to live as part of closed farming and ranching systems, farmers and ranchers will have to learn how to feed them from the land. With degraded soils, invasive species, changed rainfall patterns, and farm and rangeland often broken up into physically discontinuous parcels, this is a challenge of herculean proportions.

ECOLOGICAL MEMORY

If our goal is to restore the fertility of our soils, every available scrap of biological waste in our society should be composted, from human poop to dead bodies to food scraps to cow manure. For now, this even includes the waste from animal feedlots, trees that have been beset by beetle blight, and crop residue and agricultural waste from our chemical-industrial agriculture.

The beautiful thing about compost is that it is "feedstock independent." In other words, the process of composting does not care what you put into it, provided the source material is biological. This is because compost is an ancient process driven by microbes that always remember what to do. Those microbes and the compost they make are part of our planet's "ecological memory."

Technically speaking, "ecological memory" refers to an ecosystem's site history, soil properties, spores, seeds, stem fragments, mycorrhizae, species, populations, microbes, and other remnants and how those biological pieces may influence the composition of the replacement communities and ecosystem.[14] Ecological memory contains both biological and genetic "legacies" such as seeds, animals, and plants. In other words, it's the ability of the ecosystem to "remember" how to come back to life.

To truly bring the North American ecosystem back, we would need to bring back the large organisms, including the bison. But this is a tall order. We are now a nation of cows and freeways, cattle and cities, heifers and asphalt. Our land is subdivided into millions of parcels and our bovines are the large, domesticated, and slow kind. With our fenced-off prairies, we are unlikely to see herds of wild bison thundering across Interstate 40 anytime soon.

Today ours is a "junkyard ecosystem" with bits and pieces of species from all over the planet. It exists not because of some form of management or planning, but rather due to a complete lack thereof. In contrast, the infrastructure we value the most, including our cities, our freeways, and our Internet pipelines, are all carefully managed. For us to live well and live well into the future, the accumulation culture must come face-to-face with our ecology. We must learn to use the bits and pieces in the junkyard, as well as the strong and plentiful ecological memory of the land itself, to construct something similar to what was.

By using the techniques pioneered by regenerative farmers and "carbon cowboys," we might have a shot at a diverse, resilient ecosystem that contains many of the species that cooperatively support life, sequester carbon into the soil, and provide humans with the vast quantities of food we so desperately need. Importantly, there are also Native Americans working on this issue. People like Karlene Hunter and Mark Tilsen, Lakota Native Americans from Pine Ridge Indian Reservation who started a food company. They now make Tanka Bars. The objective of their company is to use demand for sustainably raised buffalo foods to bring back the great buffalo.

Given these leaders, and a lot of hard work, we might just be able to fill up that buffalo bank account once again.

And maybe, just maybe, with time and dedication to careful breeding and reintroduction, someday in the distant future our great-grandchildren may see herds of majestic tatanka thundering across the Great Plains again.

That's a legacy worth fighting for.

CHAPTER 6

HOME ON THE RANGE

I'm heading north on the oft-idolized Pacific Coast Highway (aka "Highway 1"). Much of its winding asphalt seems to have been fancifully sticky-taped to the edges of mountains rising up from the rocky Pacific shoreline. This is the "drive" you see in car commercials and movies. The majestic hills dotted with fauna, the ruggedness of the coast, and the edge of the road plunging down to the powerful surf all come together in a sublime experience. Even for Californians, driving "the 1," as it is affectionately called, is a revered occasion.

The town of Half Moon Bay, California, is an enclave of dot-commers who made it, people who inherited it, and people who got it some other way. Property values are astronomical, the local grocery store stocks about a thousand varieties of wine, and it's close enough to San Francisco to commute a couple of times a week (once for a meeting and once for an art show). But what really sets it apart are the jaw-dropping views of the Pacific. Rimmed by a series of stunning cliffs, the glittery bay offers a panorama worthy of a cubicle poster inscribed with a positive affirmation.

As I wind my way up the mountains just outside Half Moon Bay there's something else popping into view. It's something so commonplace that it's easily overlooked as part of the scenery. But this thing has recently become one of the most hotly contested pieces on the climate change chessboard. I'm not talking about wind turbines, cars, or solar panels.

Once again, I'm talking about cows.

A BEEF WITH BOVINES

Go to a cocktail party somewhere on the West Coast, mention the link between those . . . mmm, finger-licking-good beef sliders and "climate change," and you can be sure you'll see sparks fly. The collateral damage done by beef consumption has been vehemently debated for the better part of two and a half decades and has recently reached a fever pitch. To understand why and how we got into our on-again, off-again, love/hate relationship with beef, we have to go back to the days when disco ruled the airwaves.

It was in the 1970s when all "white" foods became unfashionable and self-proclaimed "health food" stores began to dot the suburban landscape. As the health food movement was finding its footing, a few scientists and cardiologists began to talk about a possible link between beef consumption and increased incidence of heart attacks. By the time Elvis Presley donned his famous bell-bottoms, most cattle in America had begun a steady diet of corn, something they do not naturally eat. (Corn-based cow feed would only later be linked to unhealthy fats in red meat.)

In the 1980s Bay Area–based Rainforest Action Network got Burger King to end its $35 million "rainforest beef" contract in Central America. Sting and Phil Collins soon joined in, banging the anti-beef drum. For many environmentalists, burgers became synonymous with deforestation. Lost in the frenzy, however, was an account of the subtle and complex relationships leading up to cattle roaming where trees had once grown. That cycle of rainforest destruction often begins with poverty, continues with logging, moves to cattle, transforms the land to soybean production, and ends with desertification.

In 1997, People for the Ethical Treatment of Animals (PETA) launched its "McCruelty" campaign. They organized four hundred protests targeting McDonald's restaurants in twenty-three countries, where they also used advertisements depicting gory scenes from slaughterhouses. Soon thereafter, PETA targeted Burger King and Wendy's with similar campaigns. All three restaurants kept selling burgers, but they did agree to create basic animal welfare standards—a small but definite coup for animal rights.

The early 2000s saw outbreaks of mad cow disease. Then, as if things weren't bad enough for the beef biz, Morgan Spurlock's muckraking documentary *Super Size Me* hit theaters. The film, which was nominated for an Oscar, essentially shows that fast food makes you fat. Go figure.

By 2010, it's a wonder Americans were eating any beef at all. But with the relative price of meat lower than it had ever been before, those catchy McDonald's commercials and Burger King's seductive toy-in-the-bag partnership with the hit kids' movie *Toy Story* 3, Americans couldn't resist the drive-through.

Around that time Robert Goodland and Jeff Anhang at the Worldwatch Institute had coauthored a report entitled *Livestock and Climate Change—What If the Key Actors in Climate Change Are . . . Cows, Pigs, and Chickens?* As if that question wasn't confrontational enough, right up front the authors say, "Our analysis shows that livestock and their by-products actually account for at least 32,564 million tons of CO_2e per year, or 51 percent of annual worldwide GHG emissions." CO_2e means CO_2 "equivalent," GHG emissions means greenhouse gas emissions.

Half the world's annual greenhouse gas emissions is a massive number. If the authors were correct, the climate-conscious would soon have to forgo meat.

The report, however, is based on sweeping and inaccurate assumptions. It incorrectly asserts, "Practically the only way more livestock and feed can be produced is by destroying natural forest." But in North America in the last fifty years, *much more field corn* has been produced each year on a shrinking base of farmland precisely to feed *more* livestock. Now I am not advocating we produce more corn to feed more animals, but clearly rain forest is not being chopped down in the Midwest to grow more corn.

The report weighs the greenhouse effects of cooking, packaging, and transporting meat more heavily than those from other types of food. It also counts the CO_2 from the respiration of animals (breathing out), even though the CO_2 originated in the atmosphere and was cycled into plants, which became food for livestock (and even though we had far more ruminants on Earth before industrialized agriculture). The report polarized environmentalists into two camps: people who believe livestock contribute 51 percent of the greenhouse gases and those who do not.

Adding fuel to the fire, a 2014 documentary called *Cowspiracy* was released online. Taking a page from the Worldwatch Institute report, the film claims "the meat and dairy industry produces more greenhouse gases than the exhaust of all cars, trucks, trains, boats, and planes combined." The film blames meat production to the exclusion of all other contributors for global water overuse, rain forest destruction, emissions, environmental degradation, and climate change.

Then, after years of bull bashing, the previously silent pro-cow camp got fed up. Books like *Cows Save the Planet* by Judith Schwartz, *Defending Beef: The Case for Sustainable Meat Production* by Nicolette Hahn Niman of Niman Ranch, and Allan Savory's *The Grazing Revolution* hit the shelves, painting a very different picture.

Like the anti-cow activists, these writers also eschew all forms of factory farming of animals. But instead of ending the consumption of meat altogether, they propose an ecosystemic approach to ranching and farming in which cows are central to sequestering carbon dioxide, restoring grasslands, reversing desertification, and stabilizing water supplies. These authors and the researchers they draw from assert that bovines are one of the keys to regenerative agriculture.

My question—and my motivation for my drive north—is, who's right? Are cows walking climate disasters? Or are they a critical piece of the puzzle to sequestering carbon? Or both? I'm on my way to meet a family that believes that even on the virtually waterless highlands of drought-ridden California, cows can be a tool to capture carbon and sequester it deep in the soil.

But before I meet them I need to eat breakfast.

CALLING BULLSHIT

My friend (we'll call him "Rob") is an eclectic character who owns his own organic ranch just outside Half Moon Bay. Rob is one of those people with a strong opinion on just about everything, especially when it comes to the environment. At my request of finding a place to meet that's inexpensive and fast (but not fast food) Rob invites me to the Tex-Mex café at Half Moon Bay's mostly private airport.

"Look," says Rob, wiping his mouth with a napkin after a long conversation in which I shared with him a summary of my readings on cows and climate. "I think all this cows-can-help-sequester-carbon-dioxide stuff is total and absolute bullshit." Rob's reasoning comes largely from personal experience. He tried having cows on his ranch as part of the whole "organic ecosystem" approach. He found his bovines took too much water and caused too many problems. In the end he decided the bonus of free fertilizer wasn't worth it and he sold his small herd.

As we eat our predictable diner breakfast, Rob rails passionately against the idea of using cows to restore nature. Taking a few more bites of his eggs and bacon, he says, "I'm a vegetarian sympathizer." Rob's view reflects many California conversationalists. While they aren't willing to go as far as to buy into a "cowspiracy," their collective view is that California's native lands should be left alone.

"We need organic gardens where we can and we just need to leave the California hills to go back to nature. Cows use too much water. To say this kind of thing is helping the environment—it's self-promotion, a marketing ploy by ranchers. I'm telling you, if you do a really honest study of this, you're going to come to the conclusion the cow thing is bullshit. Literally," says Rob.

He lets that sink in for a few bites. I drain my tea and ask for the bill.

Case closed, Rob perks up: "When are you going to do something about battery technology? Tesla is really crushing it right now."

We split the bill and say our good-byes.

ERIK WITH A K

Dust billows from every orifice in the vehicle as I bounce along the potholed road that winds off Highway 1 up toward the Markegard ranch. Camping outside for a week isn't my idea of a vacation, but it's the only way I am going to get to see this family of "carbon cowboys" in action. I try to focus on the scenery. Rolling hills of golden grass flecked with patches of green brush stretch out to meet the Pacific. If I were a cow, I would want to live here.

After a long and bumpy ride, I arrive at a weathered house. The Markegard family is soon out to greet me. A woman in her midthirties with hazel eyes who's wearing a cowboy hat introduces herself as Doniga. Her tall husband (also wearing a cowboy hat) with a booming voice and firm handshake is Erik, "with a K." Their four children (one boy, three girls) alternate between staring at me as if I am an alien and running after each other giggling.

"I'm sorry it's so darned dry and hot," says Erik. "You're catching us in the middle of a drought. It's really beautiful here in the spring after the rains. These hills," he motions toward the horizon, "everything here is green then. But you're here now, so why don't you come in and tell us about yourself?" (In other words, who the heck are *you* and why are you on *my* ranch?)

I follow Doniga and Erik through a gate, past an ax stuck into a log next to some chopped wood, and across a wide porch covered in chicken poop. Following the Markegards, I remove my shoes and walk through the front door. Theirs is a home with clothes, blankets, couches, and a piano. It's no *Better Homes and Gardens* magazine spread but rather a warm, friendly abode.

We sit around an old dining table in the kitchen and Doniga puts some fresh butter, some bread, and some steaming hot homemade sausages in front of me. Not one to mince words, Erik begins, "Okay, give us the spiel. Who are you and why are you here?" I swallow a big bite and give a bit of my background.

Erik is probably right to be suspicious. Having invited the *Cowspiracy* film crew into their home under what they now feel were false pretenses, the Markegards are profiled in the film as being people who love animals but kill them. The piece makes the family seem duplicitous.

Once Erik is satisfied I'm not doing a hit job, he begins to tell me his personal story. His folks had originally moved out from the Dakotas. They were poor ranchers who couldn't afford their own land. Somehow they'd fallen in with a crowd of hippies who had a commune and needed people with their hats on straight to run their cattle ranch. From there, they'd ended up managing Neil Young's ranch, which is where he grew up. He'd met Doniga because she and her friend were trespassing on Neil's land to catalogue the local mountain lions.

Before I can finish my second breakfast, Erik is on a roll. "Vegans send me hate mail saying, 'You love your kids, why don't you kill them?' " he says. "I have numerous friends who are vegans and vegetarians and I don't judge people for their diets. If their diet is for the health of their body or because they don't want to eat an animal with a face, I totally respect that. I get it." I haven't even asked Erik a question yet.

Erik continues: "We take care of our livestock. I have said numerous times and I'll say it again, I love our livestock. But people are going to eat meat. They're gonna eat meat whether the animals are mistreated or not." Erik is adamant that he is not in the livestock business to promote eating meat. Given what he sees as an inevitable culture of meat eating, his goal is to show that there's a different way of working with animals.

"There's this hate toward people that raise livestock. I'm okay with you hating people that raise livestock in a CAFO, living on concrete in their own feces. But it's not fair to hate us grass-fed ranchers," he explains.

CAFO COWS

One simply cannot drive Interstate 5 between Los Angeles and San Francisco without noticing the massive feedlots operated by Harris Ranch that stretch out toward the east. The "Ranch" has a population of one hundred thousand cattle—California's largest.

On the Web, the Harris feedlots are snarkily referred to as "Cowschwitz" (#cowschwitz on Twitter). For those weary and hungry drivers willing to overlook the smell of its animals, Harris Ranch operates a western-style restaurant and hotel. And yes, they also do weddings.

CAFO, as mentioned earlier, stands for *concentrated animal feeding operation*. If you think the name sounds bad, just type those four letters into Google and click on the Images tab. On your screen, in all of its multipixel glory, is the answer to the eternal question: "Where's the beef?" Scroll down, peruse the images, and you'll probably be having salad for lunch.

The great American Corn-CAFO complex is an evolution of the Nazi-born model of chemical calorie production. Today's CAFO meat comes courtesy of the four components of a bloated Federal Crop Insurance program, cheap plentiful petroleum, billions of pounds of toxic chemical sprays, and hundreds of millions of acres of field corn production. Like most civilizations of yesteryear, grain is the currency that underpins America. For civilizations past, that grain was emmer, kamut, or heirloom wheat. For us, the grain that runs our world is field corn.

The explosion of field corn requires massive quantities of fossil fuels for combining, transportation, fertilizer, and chemical sprays. Corn syrup and animal feed were easy ways to turn a fossil-fueled grain production machine into cash. All these fossil fuels contribute to the immense carbon footprint of the field corn and the CAFO meat for which it is grown. Fossil fuels also provide a means to centralize, control, and predict the production of meat calories.

Speaking of calories, converting energy (and water, and fossil fuels) to CAFO meat is an extremely inefficient process. The protein conversion efficiency for chicken (from grain to finished bird) is about 30 percent—meaning only 30 percent of protein from the original grain turns into chicken meat. Most of the remaining 70 percent turns into animal sewage. For pork it's 10 percent, and for beef it's an abysmal 4 percent—thus the need to grow hundreds of millions of acres of feed grain in order to create a disproportionately small quantity of CAFO meat.

Ninety-nine percent of all cows and chickens raised in the United States are "farmed" in large CAFO facilities into which they are crammed like sardines in a can. Except the animals in question are alive. CAFOs include indoor facilities for chickens and pigs and "feedlots" where cows see the sun but no vegetation.

Instead of harnessing cows to mimic the movement of bison across the plains, feedlot cows live up to their knees in rancid piles of excrement, dirt, and mud. They eat, urinate, and defecate all in the same fetid, putrid environment. Disease is rampant, which in turn necessitates the use of substantial quantities of antibiotics. Pesticides are also common in order to keep fly and mosquito populations down. Feeding a cow a steady diet of corn coats their main stomach in slime and begins the CAFO cycle of sickness and disease that is countered by drugs, which contribute to bloat, which is countered by more drugs and chemicals and so on until the cow is slaughtered.

Then hormones are used to increase "yield." These become ingredients in the animals' food supply, which in the United States can also legally contain used frying oil from restaurants (in which animals have been fried), chicken feces, and parts of other animals not used in the "meat-packing" process. Then there are the little tricks the meat industry uses to increase packing weight (aka "profit"), like injecting chickens (aka "broilers") with salt water, up to one-third their weight, before selling them. It should come as no shock that 90 percent of Americans get "too much sodium."

Erik sticks his key argument to me like a hot brand on a cowhide: people, at least people in America, don't appear to be curbing their

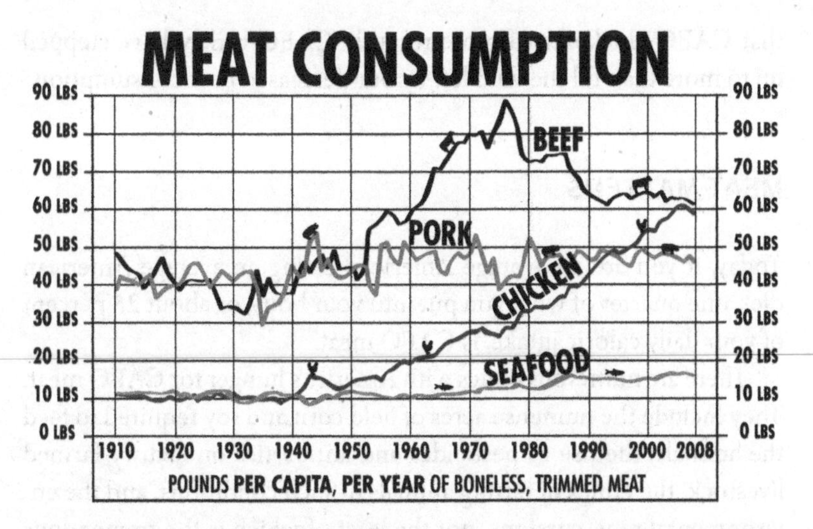

While Americans are eating less beef, they are eating more CAFO chicken and turkey.

meat consumption anytime in the near future. An anecdotal stop at the Harris Ranch restaurant confirms this. After a long drive, one finds no lack of larger than average people sitting down to dine on larger than average pieces of meat. The twenty-four-ounce bone-in rib eye is popular, as is the thirty-ounce bone-in prime rib.

In fact, apart from a dip in meat consumption after the 2008 financial crisis, Americans are on a steadily increasing "meatfest," or "meatfeast," you choose. We went from chowing down about 9.8 billion pounds of flesh in 1909 to a whopping 52.2 billion pounds in 2012. But that tidy 5x-plus increase isn't just because population has expanded. Since 1965, our per capita meat consumption increased by about 20 percent.

Today's meat consumption equates to around 213 pounds per person per year, for every man, woman, child, and infant in America (fifty pounds of beef and 163 pounds of chicken, "other chicken," pork, fish, and "other meats").[1] Some people may be pleased to learn that despite some clever ad campaigns by the beef industry over the years, beef consumption has dropped substantially since the mid-1970s. But before you conclude that America's going "veg," it's important to note that CAFO chicken and more recently CAFO turkey have stepped up to more than fill the void left by our decreased beef consumption.

MEAT MATTERS

Today, if you are an average American eating an average American diet, one quarter of what you put into your body, or about 25 percent of your daily caloric intake, is CAFO meat.

There are numerous issues with America's hunger for CAFO meat. They include the immense acres of field corn and soy required to feed the animals, the use of pesticides and antibiotics on factory-farmed livestock, the ethics of raising animals in such conditions, and the environmental repercussions, not the least of which is the tremendous

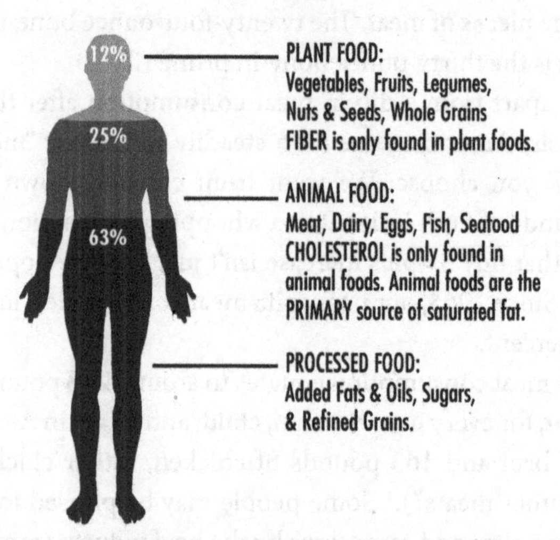

As the saying goes, "you are what you eat," and Americans are eating about one-quarter of their daily calories in the form of CAFO meat.

US OUTBEAKS OF MEAT RELATED PATHOGENS

PATHOGEN	PERCENT RANGE	OUTBREAK YEARS, SOURCE
Escherichia coli 0157	36–55%	1998–2012, IFSAC
Clostridium perfringens	16–41%	1998–2008, CDC
Staphylococcus aureues	4–19%	1998–2008, CDC
Salmonella enterica	6–13%	1998–2012, IFSAC
Shigella spp.	2.1–7.4%	1998–2008, CDC
Listeria monocytogenes	0–1%	1998–2012, IFSAC
Campylobacter	>1–1%	1998–2012, IFSAC

production of carbon dioxide across the entire supply chain. Arguably the human health effects of eating so much meat, or eating meat at all, are also a big issue.

CAFO meat contains somewhere between fifteen and thirty million pounds of antibiotics annually. With all those drugs, ground beef is often linked to outbreaks of some gnarly bacteria. In looking at some of the well-documented cases, one thing pops out: despite the antibiotics, these outbreaks occur regularly. One reason these outbreaks are endemic to meat consumption in America is that the USDA refuses to spot-test meat for lethal bacteria levels. In other words, eating a CAFO burger today is a lot like playing Russian roulette. In both cases there's nobody making sure the bullet is not in the chamber.

Ethics is a tricky subject. By the same standard that you can assert that its unethical to raise animals in confinement, you can also assert its unethical to eat them in the first place. But if Erik is right—and the last one hundred years of eating data certainly suggest he is—people will be eating animals well into the foreseeable future. At the very least, one can argue that animals we do eat ought not to be raised in horrible conditions. Americans would never stand for their pets to be housed in CAFOs. After all, cats and dogs *and cows* and other furry four-legged creatures all share the same basic biological lineage. So, why do we allow the animals we eat to live their lives in CAFOs?

The answer has to do with money.

As Steve Kay in the Meat Matters section of *BEEF* magazine

gushes, "The beef industry's contribution to the economy is colossal." A quick glance at the industry's bottom line and you can understand Mr. Kay's excitement. The USDA cites the beef industry as grossing around $60 billion a year. By the time that beef is in hamburgers and supermarket refrigerators, it's worth $95 billion. That's $35 billion to the retailers and a nice chunk to the industry.

Like other centralized resource–based industries (oil, coal, uranium, pharmaceuticals, etc.) that are moving toward duopolies, triopolies, and quadopolies, the meat industry has experienced a consolidation in recent years. Four big players now control 75 percent of the meat we eat. Theirs is a model writ large across our economy: centralize control, make production as efficient as possible, offload pesky liabilities (landholdings, insurance, management) to third parties, make waste somebody else's problem (overburdened state agencies), combine lobbying powers (into a "PAC" represented by a cushy K Street lobbying outfit), and most importantly, maximize profits.

The same mentality that brought us the Model T now brings you the great American Factory Farm. Duopolies, triopolies, and quadopolies use mechanized, centralized production in the same way monopolies do. They stifle innovation, resist scrutiny, and thwart regulation. Hence the Corn-CAFO complex, a system of powerful, centralized production whereby petrochemicals, genetically engineered grain, poisons, synthetic nitrogen, suffering, disease, and waste are the core inputs and out the other end miraculously comes cold, hard cash.

On PBS's 2002 episode of *Frontline* entitled "Modern Meat," Patrick Boyle, CEO of the American Meat Institute, said:

> America in general is a tremendous food success story. . . . We pay the lowest percentage of our per capita income on food than any country in the world. In the mid-1980s, it was about twelve percent. Today it's below nine percent. And meat, which is a large part of our diet in this country—meat and poultry—is less than two percent of our disposable income. That's a great success story.

What he failed to say was that all that "success" comes at a great cost.

In contrast to the modern miracle of cheap, mass-produced "factory meat," Doniga and Erik feel that they are fighting for something worth believing in.

A CARBON COWGIRL

I hold on for dear life as the truck slams into and out of ruts. We are climbing up a steep mountain trail. I try not to think about the vehicle toppling backward nose over end. But driving upward at a near-vertical angle seems second nature to thirty-four-year-old Doniga Markegard as she chats about the battle she and Erik are fighting against conservationists who want to take away their leased lands.

We reach the top of the hill and the truck levels off. Doniga hops out, grabs a machete, and starts poking around in the grass. I, on the other hand, need a moment before I can move from the vehicle.

Doniga grew up in Washington State romping around in nature. In high school she was placed in a nascent experimental nature immersion program with a half dozen other teenagers. At the time, the newly formed Wilderness Awareness School gave these youngsters knives and sent them out into the woods to survive. (Today the Wilderness Awareness School continues as an educational camp.)[2]

Feeling the pull of nature, Doniga went out to South Dakota to spend time on the Lakota Reservation at Pine Ridge. She was adopted as a "daughter" by the late Gilbert Walking Bull, the great-grandson of Sitting Bull. There she learned the sacred ways of the Lakota. She later found the work of Allan Savory, whose rangeland management experiments led her to the ranches of the Central California coast, Erik, and her cows.

I find Doniga a few yards from the truck examining one of the many tufts of grass. "These grasses evolved with grazers, so they co-evolved together to be mutually beneficial to each other. See all this green here in this bunchgrass?" She gingerly pulls apart the strands of golden grass to show the green strands. We are miles from the nearest water source but they are green, alive, and growing.

She continues: "And what happens if that never gets grazed? Eventually it dies off and becomes a desert. So the action of the grazing animal is what's keeping these native coastal terrace prairies alive."

Doniga admits that finding water for the cows is getting more difficult as springs and wells dry up. But she and Erik keep exacting tabs on each watershed they manage and are careful not to use more than the rainfall can replenish.

Doniga continues her explanation: "The cattle will come along and they'll eat the tops off right about to there. So that'll go in the cows' stomachs and then when the grasses get bitten off, these roots strengthen. And what is that root made out of? Carbon."

There it is—that invisible thing I am chasing: *soil-based carbon sequestration*. I still can't see it. But what I can see in this windswept California plateau, far from civilization, is substantial biological growth and not just in one tuft of grass. I look up and realize we are crouching in a veritable forest of bunchgrass, something my brain had heretofore passed off as a field of weeds.

According to Doniga, "The old accounts of California talk about beautiful green bunchgrass prairies that stood up to your waist. And the root systems of those would tap down into nutrients deep down in the soils, into water that is held deep down under the surface of the ground."

Laura Cunningham's *A State of Change: Forgotten Landscapes of California* uses "historical ecology" to reenvision how California looked before people. Cunningham's visuals give historical context to Doniga's ideal with "grassy hills, rich with bunchgrasses."

Cunningham says, "We've lost a lot of the abundance and biodiversity of the state. It must have been beautiful."[3] These descriptions conjure up something akin to the green rolling hills of Ireland, but with better weather.

Left untouched, the dry hills of California's central coast quickly cover themselves in coyote brush, an invasive, scraggly, bushy green plant that takes over grasslands, replacing animal-edible grasses with something akin to an endless inedible hedge. The problem with the "let nature go back to nature" idea is that the "native" coyote brush wasn't present in vast quantities inside the bunchgrass prairies of yesteryear. To jump-start the ecological memory of the land and begin to balance the species, you need to add back in as much of the original nature as possible, including the animals.

Doniga tells me that the managed prairie we are in isn't just grazing land for cattle. It's also home to ground-nesting birds, deer, sparrows, foxes, and other wildlife. As we poke around the grass, I watch a hawk floating overhead, searching for lunch. The squawking bird of prey certainly adds a nice touch.

She emphasizes that their cattle aren't just left to gorge themselves. The Markegards are trying to mimic natural grazing patterns by keeping their otherwise lazy bovines constantly on the move so the grass is never munched down too close to the ground. I learn that the field we are in had been left for around ninety days and is almost ready for another "mowing." Rather than a silver bullet solution to a set of system-wide ecological problems, she sees the cattle as one tool to help revitalize land systems. Without management, or with "overgrazing," Doniga is quick to point out, cattle can be highly destructive to an ecosystem and can easily reduce it to bare dirt.

Doniga Markegard explains the benefits of cow dung. (*Simon Balderas*)

Doniga shows me how the bunchgrasses on her prairie cover the earth with a kind of thatch. "You see that the soil is very hard to get to," she says as she tries to pry apart the thick cover layer of rooty grasses to get into the soil. This theme is drummed by almost everyone in the soil movement: *in a healthy soil-based agriculture system, nature covers itself. There's no bare soil.*

She summarizes the whole idea behind her rangeland management: "We're taking carbon from the atmosphere via photosynthesis into the plant, and then through the plant into the root system, and it turns into humus and builds the topsoil."

The Markegard cows are obviously not in CAFOs, but the environmental question still stands. The emissions impact of cows is a very big issue, especially for those concerned with the Earth's climate.

IT IS BULLSHIT AFTER ALL

Back on the high plateau overlooking the glistening Pacific Ocean, Doniga Markegard is crouched over a cow pie.

"It's a little dried out on top from the sun and wind but we can take a look at what comes out after the cattle graze this grassland," she says, pleased with this particular circle of crusty cow dung. Doniga turns the piece of manure over. "See, this one is just superfresh. Usually, after a day or two you start to see the dung beetles come in. And when I picked this one up the insects kind of scattered out from under it. But think about it, and in a drought like this, where is there going to be the most life? Where there is moisture. Right?"

Dung beetles are remarkable. They are one of the few nonhuman creatures to navigate using the stars of the Milky Way. There are more than five thousand known species of these small critters that collectively roll, tunnel, and dwell in dung. The dung beetle's role in agriculture is a critical one. By burying dung, which would otherwise be left out to dry, they cycle nutrients into the soil and help feed microbes, which build soil structure. Because they eat fly larvae and bring dung underground, dung beetles also serve as pest control. Maybe all this is why the ancient Egyptians revered the dung beetle. To them, the sacred image of the scarab communicated transformation, renewal, and resurrection.

Dung beetles are the uber-creatures in an entire ecosystem that begins when a cow releases its "pie" onto the ground. According to Doniga, each time one of her cows defecates it begins the process of reinvigorating the soil. "Those microbes that live in the rumen of the cattle come out and they're in this manure, this great fertilizer, which

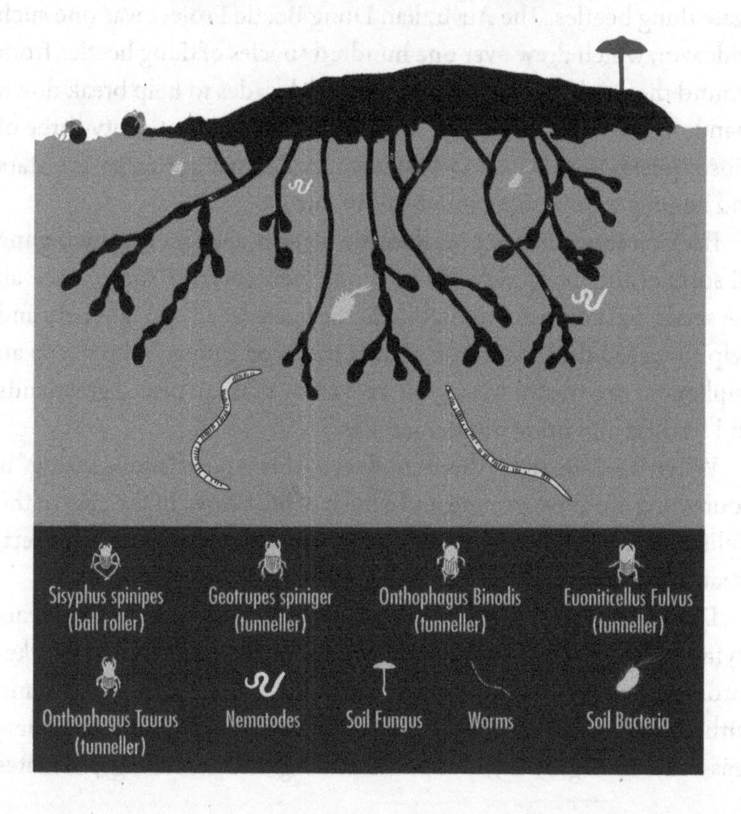

then brings more life into the soil." Her description of the bacteria in the rumen reminds me of a yeast starter culture for bread or beer. That starter culture could be key to getting soil microbes going again.

In contrast, to get rid of flies and mosquitoes, most ranchers, especially those who oversee CAFOs, use "pour-on" insecticides on their cows. These concentrated toxins do kill the insects on cows, but they also kill the microbial life in the cow's gut. Because the resulting cow manure is steeped in poison, it is often devoid of life and as a result will sit on the ground without being broken down by dung beetles and microbes. As cow manure piles up, it attracts flies, which use it as a breeding ground. The resulting explosion in disease-carrying flies necessitates more pour-on insecticides. Hence, another self-propelled destructive cycle.

All this is why far-off countries are working to cultivate and propagate dung beetles. The Australian Dung Beetle Project was one such endeavor, which drew over one hundred species of dung beetles from around the world over the course of two decades to help break down manure, increase soil fertility, and reduce flies. Today twenty-three of those species are alive and well Down Under, navigating by the stars and helping to turn dung into healthy soil.

Back on the windswept California plateau, Doniga is showing me all sorts of tiny poop details I've never seen before. "You can see all the seeds that come out undigested. These seeds will sit right here and help to reseed the grasses the animal has been eating. Who needs an implement or a tractor when you have cattle that can reseed grasslands and fertilize and bring more microbes?"

When it's done right, the aftereffect of this type of "mob grazing" is flourishing life aboveground and a flush of life below. In the case of the California drought, each piece of cow manure is a lifeboat in a desert, creating a habitat for millions of microbes.

Doniga is trying to show me what Allan Savory spent fifty years trying to figure out. "Rangeland ecology" is the basis for the Markegards' way of life: the herd walks into a prairie, disturbs the ground with its hooves just enough (but not too much), eats (stimulating new grass and root growth), poops (fertilizing and reseeding), urinates (watering and balancing pH), and moves on. All this incentivizes grasses to grow, pumping carbon down into the roots, where it can begin to be sequestered.

The ruminant needs the ground just as much as the ground needs the ruminant. This is just one of an infinite number of cycles of nature. And it is a cycle that has been happening since the dawn of the ruminant and since before humans first stood upright. But recently, we decided to stick our animals into feedlots and we broke that cycle. Now a handful of people like the Markegards are trying to fix it.

Perhaps my friend Rob had been right that it all boils down to "bullshit." Indeed, as it turns out, bullshit might just be an overlooked and critical tool in fighting climate change.

THE COW-METHANE PARADOX

Like buffalo, goats, giraffes, deer, antelope, and 150 other species of four-legged furry creatures, cows are *ruminants*. They have a four-compartment stomach, one of which is called a rumen. The rumen serves as a fermentation tank inside which specific bacteria break down the starches in the cellulose of grass and convert it into sugars. As those rumen microbes work, they create methane.

Cows emit methane gas (CH4) primarily by belching and secondarily through manure, which, in the case of CAFOs, goes into huge lagoons, where it is "managed" (i.e., "evaporated"). Methane stays in the atmosphere for about twelve years, a much shorter time than CO_2, which literally hangs out for between twenty and two hundred years. The big issue is that methane is a far more insulating gas than CO_2. Hence, the EPA calculates that, on a pound for pound basis, methane contributes twenty-five times more to the greenhouse effect than CO_2 does. So even though CAFO cows contribute a relatively small amount to the total poundage of greenhouse gases, their methane has a disproportionally large effect.

In contrast, the majority of nitrogen and CO_2 emissions attributable to livestock production are from growing the crops, mostly field corn, to feed livestock as well as the transportation of the end product (meat).[4] Since the Markegards refuse to sell their meat to anyone who is not in the Bay Area, their fossil fuel footprint is a fraction of that of conventional beef. They've also cut all ties to corn feed, making their feed supply of field grass relatively carbon neutral.

The industrial "farming" of meat does indeed contribute to our society's greenhouse gas emissions (GHGs). Aside from the singular outlier report by Robert Goodland (deceased) and Jeff Anhang at the Worldwatch Institute that claims livestock produce "51 percent of annual worldwide GHG emissions," the majority of the scientists around the world agree that while livestock is a significant GHG contributor, it is not the primary one.

Let's first discuss the global impact of livestock. According to the most recent report (2013) by the Food and Agriculture Organization of the United Nations, *Tackling Climate Change Through Livestock*, cattle, beef, pigs, and chicken account for about 7.1 gigatons of CO_2-equivalent gas emissions per year. That pencils out to about 14.5 percent of all global emissions. Various estimates, including those of the United Nations Intergovernmental Panel on Climate Change (IPCC), put the total contribution of agricultural activities worldwide (including livestock, farming, and the transportation of feed and end food products) at between 24 and 30 percent of annual greenhouse gas emissions. Thus, scientists tell us that agriculture accounts for between a quarter and a third of total global greenhouse gas emissions, and livestock account for about half that.

Now let's look at the domestic impact. According to the *1990–2014 Inventory of US Greenhouse Gas Emissions and Sinks* (published April 2016), which is produced by the EPA and draws from, among many others, Nobel Prize–winning scientists at places like Colorado State's Soil and Crop Sciences program, on a national level, around 9 percent of total greenhouse gas emissions are from agriculture, which includes crops as well as meat and food transportation. Of the US agricultural emissions, cows account for as much as 40 percent. Meaning, CAFO cows contribute 40 percent of agriculture's 9 percent of the greenhouse gas emissions in the country. This equates to around 3 or

4 percent of total US greenhouse gas emissions. It's a big number for sure, but 4 percent is small in comparison to the 31 percent of CO_2, or 29 percent of methane, produced by our petroleum-fueled transportation.[5]

The beef industry is quick to point out that grain-fed (i.e., corn-fed) beef produces *less* methane than grass-fed beef.[6] Yes, less. Their reasoning goes like this: Since they fatten the cows faster, CAFOs claim they produce less CO_2 and methane *per pound of beef produced.* However, this assertion disregards most of the greenhouse gases produced along the life cycle of CAFO meat production, from the fossil fuels to the fertilizer for the field corn, to the energy to transport that field corn to cows, to the pollution from manure lagoons. Despite the CAFO industry's claim that they are a winning ecological proposition, there is no getting around the reality that CAFOs are a significant source of GHGs and water use.

Long before humans domesticated them, there were massive herds of ruminants roaming across North and South America, Europe, Asia, and Africa. While those wild ruminants were likely not producing quite as much methane as the CAFO cows of today, the data suggests that they were producing a lot. Dr. Alexander Hristov at Penn State's Department of Dairy and Animal Science estimates that prehistoric ruminants in North America emitted 70 percent of the methane that factory-farmed animals do today.[7]

The question is, why didn't *those* ruminants alter the balance of gases in the atmosphere?

In simple terms, Earth has a number of methane-balancing mechanisms that include soil-based methanotrophic bacteria that literally eat methane, and hydroxyl (OH) radicals in the atmosphere that oxidize, or break apart, methane. In their natural state, Earth's methane "sinks" were able to absorb the methane produced by grazing ruminants. But that natural state has since been disrupted. Today increased cow methane is combined with methane from fossil fuels (particularly from fracking), methane from rice cultivation, and microbial methane emissions in the tropics.[8] The resulting overload on the planet's methane sinks has forced all this excess methane into the atmosphere.

Not a good thing when we consider the insulating properties of this particular gas.

However, there is a growing contingent of scientists who believe that, just as they can be part of the climate problem, cows can also be a powerful part of the solution. Their mantra is that it is not the cows that are the issue, it's how we manage them. The consensus among those studying regenerative agriculture is that while cows out on pasture may be methane emitters, their presence can contribute significantly to the sequestration of carbon into soil. Therefore, they argue, cows on pasture can foster a *net reduction* in overall greenhouse gas emissions.

Says Dr. Kristine Nichols, the lead scientist at the Rodale Institute: "A feedlot situation has a positive amount of greenhouse gases being emitted. In a grazing situation the data shows that we actually get greenhouses gases sequestered." While the natural state of, well, nature is in some ways long gone, there are simple steps humanity can take to balance methane emissions. Chief among them are reducing the number of cows and moving the cows that do exist out of their cages and back onto the land. This is exactly the "less cows, and more natural cows" model the Markegards are working to emulate.

RUNNING WITH THE HERD

Erik peels the truck off the 1 onto a gravel road past a farmhouse. Up ahead I can see the cattle and to our backs the sun is beginning to set.

We get out and walk toward the black-and-white cows. "We don't like to scare them if we don't have to," says Doniga. I ask her what she means. "Well," she explains, "why drive the trucks up to the them and frighten them? If we want them to move, we just call them." At this moment I realize that my image of hollering cowboys herding cattle with whips from the back of horses is based solely on reruns of *Bonanza*.

The cows affably approach us and then everyone walks together as if this is just a daily predinner stroll. As we walk Doniga makes notes.

She and Erik are checking out their cows, saying things like "One fifty-two looks like she's ready to calf," "Hey, sixty-six is still limping, we gotta check that out," and "Did you see eighty-nine? I didn't see her. She's so naughty." I ask Doniga how many of their hundreds of cows she knows by number. "Not all of them," she says, "but you tend to get to know the ones that don't stick to the herd."

It is getting dark and cold. The kids are tired and hungry. Erik calls Ranch Hand Sue, their helper, on a cell and asks her to bring one of the trucks up the hill. "We'll bring 'em the rest of the way by offering a little hay," says Erik. "We don't like 'em to eat anything but grass, but it's been a long day and sometimes you gotta motivate 'em a little."

I jump into the back of the truck with the kids, who offer handfuls of hay to the cows. As the truck bounces over the field, the cows begin to jog behind it.

Holding on to the cold steel of the truck bed, I watch the exchange between the children and the cows in what feels like slow motion. In this moment, with the faintest aura of backlight illuminating the herd moving with perfect cadence, children smiling, and the endless expanse of land and sky, this world seems very, very far away from the factory farms and CAFOs that feed our modern society.

TO MARKET! TO MARKET!

I pile into one of the Markegards' trucks, and we rumble south along the coast toward a local farmers' market. On the drive I ask Erik what is the one thing that stresses him out. He pauses for a moment, considering his answer. "Money. 'Cause it's tough. We have large bills come in but the income comes in slow in smaller amounts so it's hard to budget. When you have butcher fees and lease payments and things like that and then the money comes in at farmers' markets a hundred here and a hundred there, it's not easy."

Pescadero is a tiny enclave just off Highway 1 with surf breaks just to the north and south. It has a gas station and a couple of foodie-type sandwich shops with Wi-Fi and a weekly farmers' market in a vacant lot. By the time we get dropped off at the market it's getting close to midday, and the sun is merciless.

The market has about fifteen stands. A teenage girl croons a few folksy ballads as a handful of people mill around looking at veggies, bread, and the Markegards' stand. To say it's a slow day would be an understatement.

The Markegard Family Grass-Fed farmers' market stand has clearly been crafted by the ladies of the family. Chalked signs detailing their offerings with little Markegard cattle logos at the top adorn its sides. Their banners are hung with care. The coolers are placed just so. Martha Stewart would be proud. But in Pescadero at least, it isn't attracting many customers.

I ask Doniga what it's like for her to put in all this work and have so few customers. "I get very frustrated that I can't just pick up a phone and have meat delivered to the consumer, because I love being out on the ranch," she answers. "I love being with the animals and with nature, but there's a lot more to it. There's the marketing end and the customer relationship end. A lot of farmers, they just don't have to deal with that because it's all taken care of for them."

To ensure their meat is in fact from their own cows, and to make sure that nobody does anything untoward to their meat, Doniga, Erik, and Ranch Hand Sue literally transport their cows to everything—from grazing, to the slaughterhouse, the butcher, back to the packaging, and then to the consumer. USDA food standards prohibit them from slaughtering their own cattle so the transportation back and forth is part of life.

They personally track every cow, every cut, and every piece of meat. It's laborious, exhausting, and difficult work. While the Markegards' mistrust of the industrialized meat system may seem excessive, they feel they are basically up against the Death Star equivalent of meat.

"The industrial agriculture system is very well thought out, and a lot of systems are in place so that a farmer or rancher can sorta just plug into it," Doniga tells me. "So for a grass-fed rancher, there's no sort of system set up for us to plug into. Essentially we're creating the whole system and we're carrying the burden of it and we don't have the infrastructure, the transportation, the support mechanisms that the industrial system has."

The Markegard ladies know all their regular customers' names by heart. They catch little bits of news from one another about families and exchange recipes and cooking tips. When somebody new comes up and asks questions, the women give them honest, intelligent answers about their cattle operation and their beef.

The Markegards' workload seems endless. On Saturday, I join their oldest daughter, Lea, and Ranch Hand Sue at yet another farmers' market, in San Mateo, where they literally sprint around their stand for six hours selling frozen meat as fast as they can. Several times they can't keep up with the customers, who have to stand in line to buy their meat.

By the time the second farmers' market is winding down, I realize I'm not watching an exercise in sales and marketing so much as an education, a value exchange, and, for lack of a better word, a "connection." There's something intoxicating about getting that "special thing" from that "special person" who you know (or at least you *believe*) is truly an artisan.

The type of special thing I'm talking about doesn't come in layers of plastic packaging, and it doesn't have a big recognizable logo on it. We crave the "artisanal exchange." It is a basic and almost primal part of who we are. We want to look the maker/forger/carver in the eye and hear about the wonderful goodness that they have bestowed on that special thing, which we will then cherish or consume with extreme pleasure.

In our globalized economy this craving has started to bubble to the surface. Hence the rise of farmers' markets, of "farm to table" dinners, of craft fairs, of artisanal products sold on Etsy.com, of Fair Trade products for which producers are paid a fair price, and even of Certified Organic food. In spite of the clickable convenience and idealized ease of modernity, there are a growing number of people who want to personally know the producer of their food and how that food was produced.

GROWTH IN FARMERS' MARKETS

The number of **farmers' markets** in the United States **grew by almost 500%** in the past twenty years.

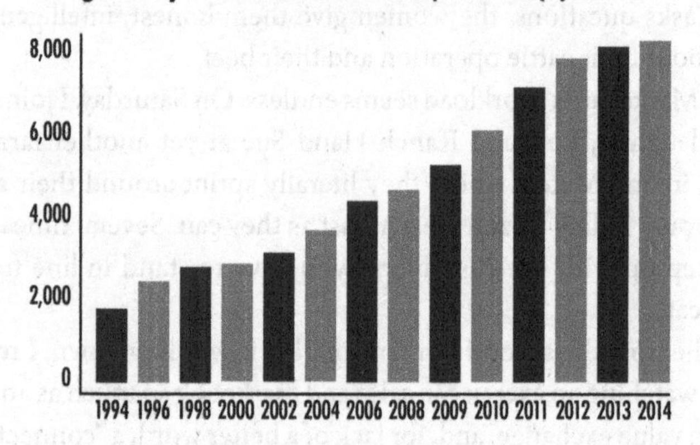

Perhaps it is a naïve fantasy to believe things can be made this way or perhaps it is a reaction to the latest iteration of our consumptive, always-marketed-to lifestyle. Perhaps. Catchphrases like conscious capitalism, the triple bottom line, the Natural Step—these are all attempts to codify, to qualify something deeper. In essence, we humans don't really want to wreck the planet and cause undo pain and suffering. The overwhelming majority of us want to do good while doing well.

Anyone who truly believes in free markets will tell you supply must eventually equal demand. Is it really so naïve, then, to think we can move away from CAFOs and move toward a healthier, regenerative model of animal husbandry that sequesters carbon into the soil? The answer likely depends as much on what we eat as what we are unwilling to put in our mouths. According to Erik and Doniga, it also depends a lot on who we buy our products from, what we know about them, and where our money goes.

THE DUST SETTLES

I join the whole Markegard family on their evening walk up the hill to watch the sun set. There's a kind of quietness that settles in after long hours of extremely hard physical work. That quietness is punctuated only by the kids as they scream and run into the fields chasing and laughing at one another. It has been less than a week since I arrived, and despite feeling the aches of having slept on the ground and the grit of dust in my teeth, there is a part of me that is enchanted with this lifestyle.

It is natural to want an easy solution to the complex challenges of rising atmospheric CO_2 and methane and a changing climate. So on one level it makes sense we are quick to blame beef as a singular cause of greenhouse gas emissions. But few folks are willing to take cold showers or stop driving cars to avoid the 29 percent of methane emissions caused by natural gas and petroleum. Fewer still are willing to stop throwing away waste to abate the 18 percent of methane emissions caused by landfills. And only a handful of really dedicated folks turn off their power completely to avoid the 10 percent of methane emissions caused by coal mining.

If we truly want to abate the emissions caused by cattle, then we need to change everything about our meat habits, from labeling to laws to the production of midwestern field corn to who we buy our meat from to what we expect to pay for it to how much we eat, all in order to change the way livestock is managed in the United States and around the world. If we are going to tackle rising CO_2 and methane emissions, less meat and more expensive meat are inevitable. Even Erik flat-out says, "I think Americans should eat less red meat. We eat too much."

Despite the Markegards' conviction and their hard work, the deck is stacked against their small operation. From a strictly rational perspective, theirs is not a winnable battle. I ask Doniga why she bothers to wake up at the crack of dawn every day to do something that is akin to bows and arrows against the lightning.

"I am going to do everything in my power to live the way I was taught to live," Doniga says, almost shaking. "And that is with the seventh generation in mind with every decision I make. So that my children and my grandchildren and my great-grandchildren seven generations into the future can be able to experience the life I've experienced and an even better one—with resources that are flourishing and clean water and clean air to breathe and trees. And I'm certainly not giving up hope, because I don't want my children to live through the mass destruction and wars and resource depletion that is a possible future."

Indeed.

CHAPTER 7

THE SOILVANGELIST

It's 5:00 A.M. Halogen lamps spit yellowish light down onto dirt-encrusted trucks, some of which still have their engines clattering. Gusts of gale-force wind burst intermittently, threatening to knock me over. Sand (or is it dust?) slaps at me as if nature herself is just getting started with her revenge. Even in the middle of summer, Kansas just before daybreak has a unique brutality.

As I do my daily run down a long narrow road into the darkness, I can sense the enormity of the space around me. On both sides of me, flat, desolate, moonlike fields stretch out to the horizon. This place is so big, so vast, it feels like it's trying to crush me. The farther away from any buildings I run, the worse the wind gets. The ground rises up to pelt me as though I'm a small object inside a sandblasting machine that is trying to rip off my skin. I am breathing in dust, covered in dust, seeing dust in front of my headlamp. I am almost exactly dead in the center of the United States and the most prevalent feature of this great land is . . . dust.

The night before, the woman at the hotel desk took one look at my driver's license and exclaimed, "Califooooornia, well, I'll be! What in the woooooorld are you doin' out here?" When I explained I was interviewing farmers about how to heal America's soil she nearly leapt over the counter to hug me. "You gotta tell the world—it's all blowing to heck, *we're dying out here!*"

As bad as my gritty run in the darkness is, it points to a much larger problem. You see, it takes nature about five hundred years to build one inch of topsoil. But the United States is losing soil ten times faster than the natural replenishment rate. (China and India are losing soil thirty to forty times faster.) Globally the situation is even bleaker. In the past forty years almost one-third of the world's cropland (1.5 billion hectares) has been abandoned due to soil degradation.[1] In case this doesn't strike you as serious, let me break it down:

No topsoil = No food. Period. End of story.

It does not matter if you add copious amounts of chemicals per acre (as we are doing), or if you genetically modify crops (as we are doing), or if you try to grow things in labs and greenhouses (as we are doing), or if you prepare your population to venture into space through time vortices to find new worlds that still have their soil (as in Christopher Nolan's *Interstellar*). Without topsoil, none of it will feed 10 billion people.

Every single thing you put in your mouth (that is food) needs soil. It is quite simply the basis for all aboveground life on our planet. And if 5 A.M. in the middle of Kansas is any indication, we are dancing on a razor's edge of blowing what little soil we have away. In a 2012 *Time* magazine article, "What If the World's Soil Runs Out?," Professor John Crawford of the University of Sydney warns, "A rough calculation of current rates of soil degradation suggests we have about 60 years of topsoil left."

The issue is not that the world's food productivity will fall off a cliff in sixty years, it's that real soil productivity will *slowly decline* as global population *rapidly increases*. Since soil is central to water retention, this also underpins constricting global freshwater supplies. In other words, this is a story of increasing poverty, increasing tension, more refugees, more conflict, and a much scarier tomorrow.

But the scariest part is that world of tomorrow is already occurring and the central issue of lower global per capita nutrition and hydration is already being overshadowed by religious wars, political disagreements, global class conflict, racial prejudice, and terrorism.

If the social implications of topsoil loss are staggering, the economics are no less severe. Organic matter facilitates plant growth and as

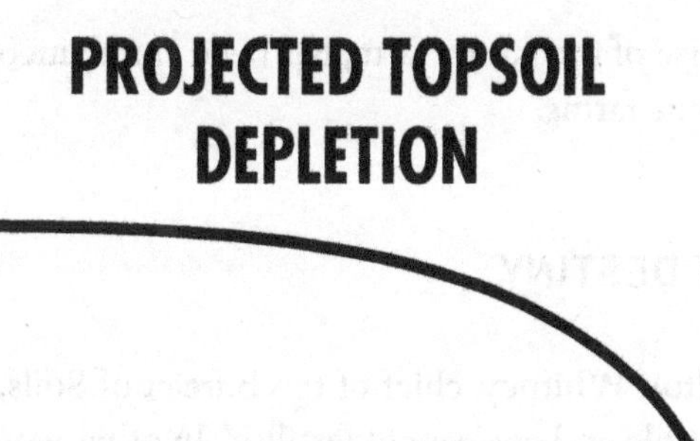

If nothing is done to stop topsoil from blowing away, we face the possibility of losing the fundamental resource that sustains life on Planet Earth.

such has an intrinsic economic value. In fact, 1 percent organic matter in one acre of soil has a market value of about $750. Since the dawn of fossil fuel use, the United States has lost between 3 and 5 percent organic matter in its cultivable soils.

The United States has around 915 million acres of land in crop and rangeland agriculture (we'll say 900 to be safe). On average, let's say we've lost 3 percent organic matter on each acre. That's just over $2 trillion worth of soil that the country has blown through in the last hundred years. We've past "peak soil" and are sliding down the slope of steady decline. The big difference between oil and soil, however, is that, at least in theory, we can build the soil back.

All this is why I'm in Kansas. I'm here to meet a man who's fighting to change the hundreds of millions of acres of fields around me. He wants to shift the consciousness of America's breadbasket. What's more, he's been hired by the US government to do it. In a strange twist of fate, the USDA, the very part of our government that laid the foundation for the Green Revolution that led to the mass commercialization of agriculture, the GMO monocrop crop insurance program, and the demise of the family farm, has been simultaneously trying to save those same farms.

MANIFEST DESTINY

In 1909, Milton Whitney, chief of the Bureau of Soils, said soil was "an inexhaustible and permanent fertility." In other words, a resource that would never (and could never) run out. Whitney was echoing a mentality that sent over one million settlers into the Great Plains and established America as the preeminent agricultural producer of the world.

That same mentality would also lead to a man-made apocalypse of biblical proportions.

Under the Homestead Act of 1862, you could get 160 acres in the Midwest for free. All you had to do was build a house and farm there for five years. That task, however, was much easier said than done. The people who ventured out into the prairies were confronted with a radically different landscape from the one that exists today. Homesteading families found themselves in "tallgrass prairies" with grasses as high as six feet. According to the Smithsonian's History Explorer, "It is said that riders on horseback could pick wildflowers without dismounting." Imagine Kansas covered in six feet of tallgrass!

The only material available from which to build houses was sod, the thick mat of grass and roots that covered the earth. Jokingly called "Nebraska marble," midwestern sod was so thick with intertwined roots that it took a plow attached to a horse or ox to "cut" it, during which it actually made a tearing sound. To appreciate the depth, complexity, and density of root structure, consider that an eighteen-by-twenty-four-inch piece of sod (approximately the size of an extra-large pizza) weighed fifty pounds. That's some dense stuff.

By 1900, over six hundred thousand land claims had been laid by sodbusting farmers and the Midwest was rapidly being transformed from grasses, flowers, natural windbreaks, and complex ecosystems into flat, open farm fields. Sodbusting transformed the land in another critical way, too. It left the soil exposed. Within one day of being cut, sod would dry out and become brittle. A casual observer could see that as this soil lost its moisture, it lost its ability to clump together. This was totally contradictory to the real estate slogan of the day that "rain follows the plow," which, as the sodbusters would soon learn, could not be further from the truth.

Rainfall in the plains region was slim, the conditions were harsh, and life was tough. But a mixture of willing immigrants, a government desirous to wrestle internal lands away from indigenous tribes, and a touch of puritanical nationalism in the form of "Manifest Destiny" was enough to spur droves of people into what had once been considered America's almost uninhabitable middle. In 1904 the US government sweetened the deal by offering 640 acres to homesteaders in western Nebraska, and in 1909 they upped the ante again by offering 320 acres in other parts of the Great Plains. World War I soon added the need for tremendous quantities of food. The race to "modernize" America's heartland (and pump as many calories out of it as possible) was on.

By 1920, five major agricultural regions had been established in America. In the South, the Cotton Belt extended across the Southeast from Texas to Florida and up to North Carolina. The Northeast was considered the "General Farming and Dairy" Belt, which extended from Missouri in the south to Minnesota in the north and as far northeast as Maine. The Midwest was the Wheat Belt, which included Montana, the Dakotas, Nebraska, and Kansas. The "Irrigated" Belt took up a good chunk of the Southwest, including Nevada, Arizona, and New Mexico, and also Idaho, Wyoming, Utah, and Colorado. Finally, the "Fruit" Belt encompassed the West Coast, including Washington, Oregon, and California.[2] Along with the Transcontinental Railroad (which would later be supplemented by freeways), these belts formed the basis for the very agricultural system America has to this day.

Between 1920 and 1930, a second land rush in Texas, Oklahoma, and Kansas added almost two hundred thousand people to the region. That land was flat, free of stones and stumps, and well suited for machines. Gasoline-powered tractors, disc plows that "pulverized the surface soil," and combines that could harvest five hundred acres

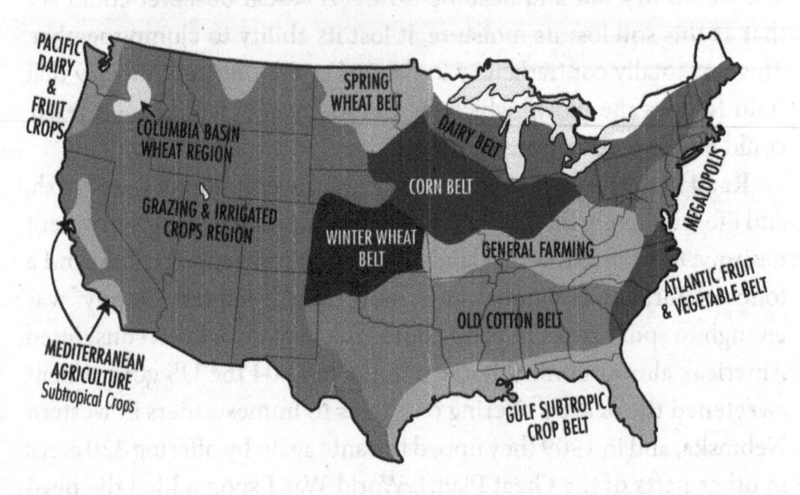

The agricultural regions that were established after World War II are largely still in place today.

in two weeks soon burned and churned their way across the plains. By 1930, over 75 percent of farmers in the Winter Wheat Belt had machines. Where tallgrasses once stood was now an awe-inspiring network of fossil fuel–powered food cultivation.[3]

Man had finally conquered nature. Or so it seemed.

BLACK SUNDAY

On March 21, 1935, the clock was counting down for Hugh Bennett. The country was in crisis, and he was one of the few people who understood how to fix it. But his agency was out of money, his speech was almost over, and Congress wasn't budging. He needed a miracle.

Having grown up in North Carolina on a cotton farm, Bennett had been introduced to the importance of soil at a young age. This would lead him to spend the majority of his career looking at dirt. As a field researcher for the Bureau of Soils under the Department of Agriculture Bennett had surveyed soils across the country as well as overseas. His conclusion was "soil wastage" had become a serious threat to the United States. He wanted to reevaluate national farming practices, especially the practice of tilling.

As tends to happen with any large-scale national crisis, America was experiencing not one but two major catastrophes. The economic crisis (precipitated by the 1929 stock market crash) was being compounded by massive problems in the Midwest.

By 1931, the rain had stopped, the plains had dried out, and fierce winds ripped across the middle of the country. Then, like something out of a horror movie, came the horrific storms. Thundering black clouds of dirt began to roll across the plains. The topsoil that had taken nature a thousand years to build was vanishing before farmers' very eyes.

From Texas to New York, walls of dirt soon blanketed America, in many cases blacking out the sun and reducing visibility to a few feet. The dust storms increased from 14 major storms in 1932 to 180 in 1933.[4] Choked by dust, animals lay dead in fields. People left the Midwest in droves for California. Those who stayed fought for survival. And many of them continued to plow.

In 1933, Congress created the Soil Erosion Service, putting the closest thing they had to a soil expert—Hugh Bennett—at its helm. By the time Bennett stood in front of a congressional subcommittee, the problems of soil erosion and dust storms had reached epic levels. To combat the problem, Bennett needed more funding, more manpower, and more authority but his Washington counterparts had mired his small agency in an administrative spiderweb of bureaucratic immobility.

Before taking the podium, Bennett had received telegrams from his assistants in the Midwest that a massive dust storm was approaching the nation's capital. If he could time it correctly, his speech might just coincide with the storm's arrival. He pontificated, he took his time, he dragged each talking point out, all the while hoping that disaster would strike. To everyone's surprise Hugh Bennett's storm arrived just in time.

According to Wellington Brink's book *Big Hugh: The Father of Soil Conservation*:

> The group gathered at a window. The dust storm for which Hugh Bennett had been waiting rolled in like a vast steel-town pall, thick and repulsive. The skies took on a copper color. The sun went into hiding. The air became heavy with grit. Government's most spectacular showman had laid the stage well. All day, step by step, he had built his drama, paced it slowly, risked possible failure with his interminable reports, while he prayed for Nature to hurry up a proper denouement.

According to a program produced by PBS member station WETA, "When the sky went dark outside the windows of the hearing, one member of the committee reportedly noted, 'It is getting dark. Perhaps a rainstorm is brewing.' 'Perhaps it is dust,' another ventured. 'I think you are correct,' Bennett agreed. 'It does look like dust.' "

The fact that midwestern topsoil was piling up on the streets of Washington, DC, and as much as a foot thick on ships in New York Harbor, was a serious issue. In response to the "Dust Bowl," President Franklin Delano Roosevelt signed into action a bill that would create an entity called the Soil Conservation Service as part of the Department of Agriculture. Thanks in part to Bennett's well-timed testimony, that entity would have permanent funding and a mandate to save the nation's soils.

Bennett and his successors would work tirelessly to establish windbreaks and other conservation measures across the farming and ranching belts of the country, thereby considerably reducing soil erosion. Believing the nightmare was over, America went back to the business of "producing" food. By 1950, a million more acres were added to the land that had been under plow during the Dust Bowl. Mechanization was soon met with heavy use of synthetic fertilizers and pesticides. America's post–World War II "industrial" agriculture was off to the races.

Bennett died in 1960, long before he could finish his mission of conserving America's soils. The Soil Conservation Service was eventually renamed the Natural Resources Conservation Service (NRCS). While its mission would remain the same and it would eventually put thousands of men and women on the ground across America in an attempt to safeguard the nation's soils, its powers would be heavily limited by the USDA and by all that happens in Washington. As a result, instead of the dark storms of Bennett's era the Midwest of today is plagued by a chronic, steady depletion of the soil from wind, tillage, and overirrigation.

They say that a frog dropped into boiling water will leap out, but one that is boiled slowly will stay to die. Luckily for America, there's at least one man who's fighting to make the frog leap before it's too late.

FLATLAND

Breakfast is the standard low-end hotel mix of rehydrated eggs, rehydrated oatmeal, and eat-at-your-own-risk mystery meat. I opt for the oatmeal. While working on a project about healthy soil and healthy food, many of my meals are on par with prison food. After months on the road, the Standard American Diet with its nutrientless calories is starting to take its toll on my body and mind.

"That's bad news," I think, looking at the GPS as my vehicle plows a straight line through the early Kansas morning. I underestimated the travel time from my hotel to my next location. The place I am going is literally two and a half hours from the nearest hotel.

If you want to get a sense of how America's soils are doing, take a drive across Kansas. For hours one passes through a stark, flat, endless sea of bare, dry, cracked earth. Occasional fields of crops punctuate the barren ground.

Those circles of crops you see from the comfort of your pressurized airplane cabin are created when a "pivot" (a long metal irrigation trellis with wheels) moves from a central point around and around in a circle spraying water. (Think of holding down one end of a pen with one finger while moving the other end with your other hand in a circle.) The pivot provides the water needed for a given crop (mostly soybeans or corn).

Looking out the car window across Kansas, one can see a pattern. Pivots and crops, barren land, wind; pivots and crops, barren land, wind. Rinse, lather, repeat. This is how and where America gets the bulk of its calories. And this is also how we lose our soil.

THE DUDE WITH A BELT-CLIP CELL PHONE

"Okay, I want you guys to start off by giving me a gooooood Kansas 'Good morning!' " says Ray Archuleta, who is standing in a steel barn in the middle of America looking at thirty-five or so graying farmers. The mostly overalled men sitting on folding chairs in neat rows oblige with a dull "Good morning." (I feel like we are either going to speak in tongues or buy Tupperware.)

Ray Archuleta, speckled gray hair, off-brand polo shirt, jeans, is in his fifties. Like a six-shooter at the ready, his cell phone is clipped to his belt. For Ray, that's probably a good thing. The man seems to talk at an incessant speed to crowds, to people on the phone, to just about anyone who'll listen. You can tell right away this is a guy who has something pressing to say, even if it doesn't always come out exactly right.

I am here to watch him do what he spends the majority of his life doing, a practiced lecture/dog-and-pony show for farmers (aka "producers") whose average age is sixty-five and whose average interest level in what he's saying is at best lukewarm. But none of that seems to bother Ray, whose official title is Conservation Agronomist for the NRCS. In some ways, Ray is a modern-day Hugh Bennett.

"We're gonna show you how to farm like nature. And the more you farm like nature, the more we can reduce your inputs and the more money you can make," he says, leading with a combo mission statement–sales hook. Ray explains that he collects soil samples from all over America and stores them in his garage. He has brought some samples to share with the group, as well as some soil from a local farm.

He has the men gather around a folding table placed at the front of the room. Two large, clear plastic cylinders full of clean water sit neatly on the table along with some soil samples. It is obvious that this demonstration will involve the soil going into the water. What is less obvious is what Ray is about to reveal.

"This is called the *soil aggregate stability test* or *slake aggregate stability test*," says Ray, emphasizing this "teachable moment." He explains that this science has been around since the 1930s and is something anyone can do on his or her farm, ranch, or home. Ray begins with two clumps of soil from farms close to where he now lives in North Carolina.

Clump number one is from a farm that has been "no-till" for forty years. That farmer also plants a multispecies cover crop in between planting his cash crops (so his land is never fallow and his soil is never bare); he uses no synthetic nitrogen, no phosphorus, and only one herbicide to reduce stubborn weeds. Ray says that typical organic matter in that area is around 3 percent, but this farmer had achieved 6.5 percent organic matter in his soils. Clump number two is from a conventional farm in the same area that uses heavy tillage, chemicals, pesticides, and fertilizers. In other words, "the usual."

It's time to drop them into the water, but Ray is savoring the suspense. "Just look at the color difference. *Look at that.* I can see the difference in these fields from thirty thousand feet," he prompts. The men lean in, eyeing the soil clumps. No-till clump number one is a dark, rich chocolatey color. Heavy-till clump number two is dry and orangey-looking.

"Now we're gonna drop them in and watch what happens," says Ray, prompting his volunteers to drop the soil clumps into their respective water cylinders.

By this point, Ray is in his element, a master on the stage. "So when he drops 'em in, the water is gonna rush in to fill the pore spaces. As the water is rushing in to fill the pore spaces, it should hold its integrity. The *biotic glues,* the organic mineral complexes, should be able to hold the clay and sand particles intact and maintain the pores." To make his point, Ray is motioning to clump number one (the no-till, chocolate soil clump), which, even though it was immersed in water, is holding together nicely.

The second (heavy-till) clump isn't doing so well inside its water tube. "We don't want the soil to fall apart. If it starts to fall apart, to *slake*, that means the pores are collapsing. If the pores collapse, no porosity. No porosity, no infiltration. Then you get crusting, and the majority of that expensive water that you pump runs off and evaporates. We don't want it to evaporate. We want it to go into the soil. Right, guys?" This is, of course, a rhetorical question. The soil in clump number two is quickly falling apart, turning the water murky. "This is just one indicator of soil health, but it's an important one," explains Ray.

TO TILL OR NOT TO TILL

Over the course of the next five days, I watch Ray give this demonstration as the opening act in his "soil show" in hotel conference centers, barns, and government extension offices across Kansas. He sometimes uses a soil sample from a certified organic farm that employs heavy tillage, emphasizing that being "organic with heavy tillage" is possibly more destructive to the soil than being "conventional with zero tillage."

At times, it seems like Ray is sending a mixed message with his tilled-organic versus no-till conventional "soil clumps." Especially when he says things like, "Look at that soil from the organic farm—*the bastion of sustainability!*" as the organic soil disintegrates into the water. As Ray explains it, "Our soils can handle acute stresses, but they cannot handle chronic stresses. So the more tillage, the more chemical and biological disturbances, the more the soil biology suffers.

"Organic systems still use huge amounts of physical stress, especially tillage. I tell my organic producers they're not sustainable because they're using huge amounts of tillage and they're still very heavily dependent on diesel to move the manures and the composts. Some of the most degraded farms I've seen in California were organic farms. They use huge amounts of tillage. Their soil was incredibly degraded. One rain event there causes massive erosion and loss of organic matter," Ray says.

"Industrial organic," as Michael Pollan calls it in *The Omnivore's*

Dilemma, is the practice of performing certified organic agriculture on a similar scale as conventional farming. To grow massive quantities of monocrops in the same soil with little or no crop rotation year after year means you have to deal with weeds. There are certified organic, biologically derived herbicide sprays, but they don't have the killing power of nonorganic weed killers (which are generally toxic and tend to remain in the soil). Nor do certified organic herbicides have the effectiveness of handpicking. For many large-scale certified organic farmers this leaves only one option to get rid of weeds—to till them into the soil. Hence, the "organic dilemma" of no chemicals but lots of tilling.

"People equate organic with being very good ecological stewards. That's not always the case." Ray pauses, adding a little weight to his point. "But one thing I will tell you about the organic community that I love is that once you teach them more about how the soil works and how it functions, they are very quick to respond and change."

Unfortunately, there's plenty of evidence to support Ray's claims. Across the Midwest and throughout California one can find heavy-till organic farms that have dry, compacted, brittle soils and many no-till conventional farms that have rich, dark, thick soils. The biggest difference in soil health in these extreme examples is not the use of chemicals versus no chemicals. Instead, as Ray asserts, soil health is directly related to whether or not the farmer tills the soil.

When it comes to tillage, Ray is merciless on his audiences, asking questions directly to his producers, who respond mostly with blank stares. He holds up the cylinder filled with murky water, its soil clump all but dissolved, and asks the audience, "So what's going on here? What does tillage do? It's breaking up the organic matter. How does it do that? *How does tillage do that?!*" He stares at the audience for a long time awaiting a response. Terrified, nobody answers.

But Ray doesn't expect them to, so he gives them the answer. "There's a really powerful tool for doing that right across the other side of this shed. It's called a disc. The highway department uses a disc for a very special purpose. It wants to destroy the pore spaces, and wants to oxidize the organic matter. The tilling disc is the most destructive thing we have out there. Why? . . . What is it doing? . . . Come on, producers, *why*!?" Again, for fear of death, nobody answers.

"You should know this. We should know this. All of us should have been taught this. We're going broke because we don't understand this! When you run that disc, and you do all that surface disturbance, guess what you do?" This time he isn't waiting for a response. "You wake up these bacteria called r-strategists bacteria. They're in all soils. They're opportunistic bacteria. They feed on rich carbon substrings. Once they eat the carbon, they die off and release nutrients. So when you run that vertical disc that you spent $50,000 on, when you do surface disturbance, you oxidize the organic matter off."

John Deere's website advertises its disc harrows with the motto "as low-cost as they are groundbreaking." (Pun, I'm sure, intended.) Basically, these tools are large circular plates that trail behind a tractor, diving into the soil and ripping it open. John Deere's "Frontier Utility Disc Harrow" is a "handy tool" that "does it all, from spring and fall tillage, to mechanical control of weeds." My personal favorite is the TM51, which features (and I quote) up to nine inches of "aggressive penetration." Don't like that one? Another option might be the "Drawn Offset Discs" for "punishing tillage jobs." If ever there was a collection of modern industrial farming metaphors for Mother Nature being violated by Man, this website is it.

Back in the field, Ray is doing his best to explain a fundamental and extremely important concept, one that if truly understood could overturn modern agriculture entirely. In essence, running a plow over soil and "breaking" the soil, the very thing that man has been doing in agriculture since the dawn of human civilization, destroys the organic matter (which contains life-giving carbon and microorganisms). Put more simply, tilling (aka plowing) makes the soil cannibalize itself.

The complex version of the story (which a handful of farmers in Kansas who have spent half their lives tilling soil are trying to wrap their minds around) is that tilling exposes certain strains of bacteria to the oxygen in the atmosphere. Once these little critters have access to air, they binge-eat the organic matter in the soil. They oxidize (or burn) the organic matter off, stripping out the stable, carbon-laden molecules and releasing substantial quantities of CO_2 into the atmosphere. This is the *opposite* of carbon sequestration. If putting carbon back into the soil is the objective, then tilling is akin to punching the ship full of holes before sending it out to sea.

Ray is doing his darndest to reverse an estimated ten thousand years of agricultural thinking, and it's not going well. Of the seven hundred e-mails the NRCS local office sent out about Ray's field day, thirty-five farmers showed up. I ask Ray how many of his audience he might convert. "If I get one in ten, I'm a happy boy," he says. "Because farmers watch their neighbors, and that guy is gonna affect his whole community." Still, the odds of Ray shifting the 915 million acres of US production with one dude per thirty-five, per lecture, per day, per barn somewhere in the middle of nowhere seem awfully slim.

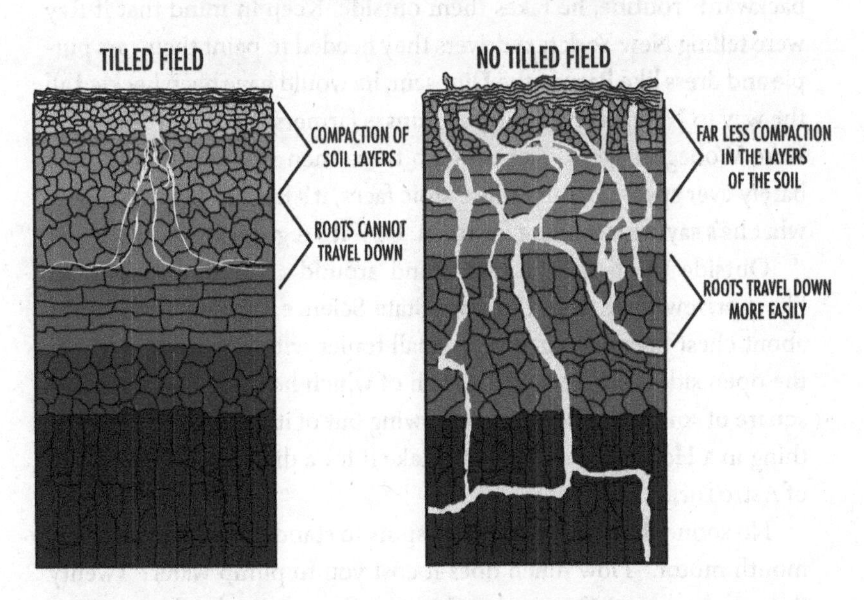

Farmers till to remove weeds and level the soil, but tillage compacts the lower layers of the soil, making it difficult for plant roots and water to penetrate.

The issue that Ray is coming up against is that the entire farming infrastructure of the Midwest is based on tilling the soil (and selling machines and chemicals, which we will get to in a minute). Soil is tilled to aerate it, to mechanically destroy weeds, to even it out so machines can pass easily over it, to loosen it up, and mix all the "good stuff" (crop residue, chemicals, whatever) into it so it's homogeneous.

As Ray points out, tillage creates a self-perpetuating cycle. As you till, you kill soil life, which then destroys the glues that make the soil porous, which means that the soil can't absorb water, which means it needs to be tilled to add "air," which creates erosion, and so on. In other words, the more you till, the more you need to till.

LET IT RAIN

Once Ray warms up his audience with his "drop the soil clumps in water" and "tell them their fundamental understanding of farming is backward" routine, he takes them outside. Keep in mind that if Ray were telling New York taxi drivers they needed to paint their cars purple and dress like Barney the Dinosaur, he would have been heckled all the way to New Jersey. But these Kansas farmers are good Protestant, Lake Wobegon types. Not only do these men not heckle Ray, they barely ever speak. Behind those stoic faces, it's hard to figure out how what he's saying is landing for them. Or if it's registering at all.

Outside the barn, the men stand around a machine that looks like a prizewinner from the Iowa State Science Fair. It's a white box about chest high mounted on a small trailer with one open side. On the open side sit three shelves, each of which have a 1 foot by 1 foot square of soil with a bit of grass growing out of it. If you passed by the thing in a Home Depot, you'd mistake it for a display on the benefits of AstroTurf.

No sooner have the men found spots to stand than Ray revs up his mouth motor. "How much does it cost you to pump water? Twenty thousand an acre? Guess what, I can cut your water by fifty percent. *Fifty percent!* That's because most of that expensive water coming out of your pivots is evaporating or running off. You're wasting money. But I want to put money back in your pocket. Okay, my friend Dale here is going to tell you about this rain simulator. Take it away, Dale."

Like a magician's assistant, Dale (also an NRCS field agent) is well mannered and shorter and rounder than Ray. Dale takes center stage, dutifully explaining that rain in the Midwest generally comes in one of two varieties: a short shower or a large thunderstorm. The rain simulator, he says, simulates the latter. It's designed to show how various soils react to a "heavy rainfall event," as he calls it. Hence there is a sprinkler contraption above each square of soil.

Dale points out another important part of the machine—a "runoff" jug and an "infiltration" jug for catching the water from each soil square. "There's holes underneath these soils so anything that goes through the soil profile will go into this back jug," he says, pointing to the infiltration jug. "Anything that runs off, obviously, will go into the jug in the front," says Dale, pointing to the runoff jug.

The three soil samples have been pulled locally. One is from a conventional farm with heavy tillage, another is from some grassy rangeland with no tillage, and the third is from a no-till, cover-crop farm. Dale turns on the machine and the men watch as the water gushes down and filters through the soil samples and into the jugs.

The results are both predictable and unsettling. The soil from the conventional farm fairs poorly. The runoff jug quickly turns brown, indicating that plenty of soil was being carried away by the rain. The infiltration jug also turns brown. In comparison, the water in the jugs from the other two soil samples seems clear. Even more curious, both of the other soils seem to be soaking up more water, meaning there's less water collecting in their jugs.

Each time Dale explains the machine, Ray jumps in right behind him with a string of zingers. Ray points to the dry, conventionally farmed soil sample. "Everybody get close because we're gonna show the early stages of desertification," he says.

"When you have bare ground, you have heat coming into that soil, and it's called *sensible heat*. Sensible heat means you can sense it. It raises the soil temperatures incredibly. At 113 degrees, you shut down the bacteria. You shut down enzyme activity. Enzymes are the catalyst for nutrient cycling. Then there's another heat called *latent heat.* It's where the water evaporates. This will evaporate huge amounts of water," says Ray, laying the framework for a much larger explanation that is about to come.

By now several of the producers are craning their necks trying to get a better look at the patch of dry dirt. Ray has their attention, and he knows it. "Which attracts the rain?" he asks. "This one or this one?" He wants them to choose between the conventionally tilled dirt and the rich, grassy soil. "That does," he says, pointing like a rod to the grass-covered soil.

"Why?" he asks, pushing his audience to think about something they've seen their whole lives but had perhaps not internalized. "When you have too much sensible heat coming off, heat vortexes go up and actually push the clouds away. But when water comes out through a living plant you're sending humidity out, it's cooler, and it contributes to rain. Forty percent of our rain comes from inland." This is one of Ray's big points. Just like tilling begets more tilling, dry, arid land contributes to desertification.

"So guess what? Producers, you created your own drought here. All that fallow land is contributing to your drought," asserts Ray. "Remember, this was a prairie. It was covered twenty-four seven. You want the moisture to come out of the plant, then you increase the humidity in the air and then when it gets a little colder, it rains." In one fell swoop, Ray has delivered another modern farming paradigm–shattering realization.

Ray is convinced the small water cycles are being disrupted around the world by modern farming practices. "We create our own deserts," he repeats over and over to crowds big and small across the Midwest. If he is correct, the power to worsen, halt, or reverse desertification has never been so close at hand.

FIELD DAY

Like a sheepdog herding cattle, Ray gathers the men and steers them out toward a field several hundred yards from the barn. This is the third piece to his program—getting into the soil. He picks a spot inside a leafy green crop where everyone can stand around.

Before half the men amble up, he launches into the next chapter in his soil sermon. If Ray had spent his morning boxing, he's now removing his gloves and going for the jugular. "Guys," he says with a loud, booming voice only partially justified by being outside, "we farm and we don't understand the ecological principles of how our soils work and we're going broke over it!" Ray plunges the shovel into the ground and unearths a mound of soil.

He kneels down on the ground and exclaims, "Look at that soil!" as if he is holding the magic of the universe in his hands. The men look intently as if enlightenment is somehow possible through squinting. "Get closer; here—look at that. When I dig a shovelful I want to see this 'cottage cheese,' that's what I call it." Ray is referring to the dense mass of roots and soil aggregates—the "clumping" of the soil.

"Scientists call this the *aggregate sphere*," Ray explains. "Inside the aggregates are pores. If I've got lots of aggregates, I've got big pores, not little pores, *but big pores*. Those big pores get destroyed with tillage." Like a sponge full of small holes, the pores inside the soil are what hold water. Without the pores, "infiltration," as the farmers call it, is low and soil will not retain moisture.

The more water that infiltrates the less a farmer has to water his crop. Since water is a major expense, this is important. Ray explains that "in good prairie and forest systems, we'll have infiltration rates of fifty to eighty inches per hour."

The velocity at which water enters the soil is measured in inches per hour. Most agricultural soils today can only hold an inch or less of water. In contrast, for every hour of rain the untouched, untilled soils of the Midwest once retained as much as eighty inches of water. If this seems like an incredibly large volume of water, consider that the largest ecosystem outside of Earth's oceans is not above ground, but beneath it, and it's a very wet ecosystem.

According to Ray, "If we were all right now to be able to scrape down, we would see billions of bacteria. We would see this little *mesofauna*. It is a living ecosystem. Let me give you an example. A good, healthy soil should have the weight of a cow and a calf or even an elephant of microbes under every acre. That's anywhere from twenty-five hundred pounds per acre all the way up to ten thousand pounds of organisms per acre." The unseen world that the producers are all squinting at seems so immensely vast, it is impossible to fully imagine.

As if Ray's case were not difficult enough, he takes his audience another step down the unseen microbe rabbit hole. "Those soil microbes, they swim in water, they don't have legs—they're aquatic creatures," he says, explaining that soil is essentially a semiaqueous solution. "So if you have no water, you've got no nutrient cycling. I don't care how much fertilizer you put out there." Blank stares from his crowd indicate either a lack of understanding or a slow, painful realization.

"The microbes and a majority of the microbiology are in the top two inches," explains Ray. "So if you don't get this first two inches right, the others don't even matter." He was basically telling these poor dudes they were creating a desert and starving the life that would otherwise live inside their water-based soils—the very life that could, if nurtured, turn their farm operations into profitable businesses. Not exactly good news at 11 A.M. on a Tuesday.

"Your soil is made out of these fusions of 'cottage cheese' created by fungus, earthworms, bacteria, polysaccharides, and the biotic glues. If I don't see 'cottage cheese' in your soils, guess what it's telling me? You're tilling too much, and it's starving to death." For Ray, the soil is a character made of many tiny characters, each with its own needs and gifts.

"Next thing the soil should show me is at least three or four earthworms per shovelful," explains Ray as a worm falls out of the soil and scurries back into the ground. "Three to four earthworms in a shovelful equals between 850,000 and a million and a half earthworms per acre. You don't have to put diesel in them and they cycle nutrients for you." This is news for a group of people who are the targets of substantial amounts of advertising from the companies that make the very chemical sprays that kill earthworms (and just about everything else).

Charles Darwin was said to have been obsessed with earthworms and to have studied them for thirty-nine years. With 2,700 varieties, these cold-blooded hermaphrodites greatly range in size. (The longest earthworm found was in South Africa at twenty-two feet.) But the reason that Ray (and many other soilophiles) loves earthworms is because of what they leave behind. Earthworms tunnel through the soil, aerating it, creating avenues for water to penetrate and in their wake leave a steady stream of "castings." The castings are really where it's at. Compared to the soil in which worms live, their castings are five times richer in nitrogen, seven times richer in phosphates, and eleven times richer in potassium. What's more, a happy worm living in good soil can produce ten pounds of castings *per year*! That is a lot of microbe- and plant-ready compost.

"The other thing I want to see is 'armor' on the soil," says Ray, who is using another Rayism to describe the thick thatch layer that covers untilled earth. It's that woven-like material you have to scratch through in undisturbed ecosystems to even get to the soil.

"That's called the *detritus sphere*," Ray explains. "This armor provides weed suppression. I want my covers growing right out of that armor." Ray is proposing that the ground should be covered in thatch, and into that thatch farmers should plant a cover crop made of multiple plant species. The two keys to no-till agriculture, explains Ray, are not tilling the soil and using cover crops, instead of leaving the land bare between cash crops.

As it turns out, the species for that cover crop are important because of what their roots do. "So when we do a multispecies cover crop, I want the ones that are fibrous with deep taproots because I want carbon pushed as far as I can into that system," says Ray. "So if you look at the height of your corn, it should be pretty much the mirror image of what should be down in the ground seeking nutrients and water. And so where you have roots, you have microbes grazing right at the edge."

Plants actually "leak" carbon through root exudates, which is basically "carbon juice" that's often in the form of sugars. Those sugars are consumed and transformed by another set of underground creatures, including nematodes and fungi.

The other key role of cover crops? Done correctly, they can "fix" tremendous quantities of nitrogen into the soil. Because the atmosphere is 80 percent nitrogen, there are approximately thirty-five thousand tons of nitrogen over each acre of land. Ray explains that planting legumes as part of that cover crop "mix" ensures that not just carbon but also nitrogen is being sent down the roots of those plants and carried into the soil by microbes.

Says Ray, "It takes copious amounts of natural gas to strip that nitrogen molecule and make chemical fertilizer. To me, that's an incredible waste when I can just 'grow' my nitrogen. Legumes and microbes do a fantastic job of producing their own nitrogen." Used wisely, he says, cover crops can completely eliminate a farmer's need for synthetic nitrogen fertilizer. And by eliminating one of the most expensive and polluting chemical inputs of modern agriculture, farmers can save a lot of money.

Like a plow going over and over the same piece of land, Ray is slowly etching his point into his audience: Water, carbon, and nutrients follow the same pathway and cycle. If you can create microbe-friendly habitats inside the soil, that's where you'll find the nutrients, water, and carbon. If not, you've got to constantly add those things using pumps, sprays, synthetics, diesel, tractors, machines, and lots of money.

The sun is now high above the men in the field and sweat is the predominant facial feature on his audience. But Ray isn't done. He takes a few moments to drive his last nail in. "So every time you mismanage carbon in this environment you're gonna pay, you're gonna pay dearly. If you want to farm and make a lot of money, emulate nature. If you want to go broke, watch your neighbor."

RAY'S WORLD

It is the third night of the Kansas-Archuleta tour de force and I have followed Ray to one of those generic freeway exit hotels that look eerily similar to a Holiday Inn, right down to the colors on the marquee. The irony of an emulation of a Holiday Inn in Nowheresville, America, is only elevated by the local food choices. Or should I say choice. The only restaurant within walking distance is Applebee's.

Let me first praise Applebee's as the neon mecca of cheap late-night margaritas and sympathetic hostesses that transformed so many of my otherwise drudgerous graduate school nights into nacho-filled party-time memories. Okay, blurry memories, but memories nonetheless.

When you're sober, however, and you're traveling around America learning about nutrient density and microbes and carbon cycling, this particular eatery becomes just another logo slapped onto the endless calorie conveyor belt that is our fast-food nation. Yes, the margaritas are still inexpensive but so many of the savory menu items are ultimately derived from corn and soy, most of which is grown in places like Kansas with chemicals and irrigation pivots and soil blowing away and the rest of it.

By the time I drag myself through the door, Ray and his entourage are on Miller Time. I saddle up, order some cardboard-like fare, and prepare for some well-lubricated storytelling.

Ray grew up in New Mexico, where his family has been for five hundred years. There's even a county named after his family just across the state line in Colorado called Archuleta County. His father told him when he became a man he had to do two things: work hard and not marry a cousin. Ray had gone to college to study agronomy, volunteered with the Peace Corps, married a farm girl from Missouri, and together they had two daughters. He's been working for the US government for the better part of three decades. In the process he also fulfilled both his father's requests.

Ray explains that his average day begins around 5 A.M. He spends the first hour getting dressed, answering phone calls, and responding to urgent e-mails from "producers all over the country." Most of his year is spent traveling. For him, it's a mission. I ask him how it started.

"I've been doing conservation for thirty years, and everywhere I work, we're still having the same problem—producers are losing the farm," Ray tells me. Indeed, in that same period of time that Ray has been working to save farms, the United States lost approximately 350,000 farms, just over 10,000 farms a year.

"Modern agriculture was built on the wrong premise. It was based on the premise 'let's force it, let's manipulate it, let's genetically change it.' " I ask him why the majority of farmers appear oblivious to that "wrong premise." "Our chemical fertilizers mask the problem of degraded soils," he explains. "Farmers will say, 'But Ray, I'm growing two hundred fifty to three hundred bushels of corn.' And I say to them, 'But are you making money?' "

According to the USDA's own statistics, the state of farm income in the United States is nothing short of dire. In 2011 net farm income hovered around $114 billion. But by 2016 it took a nosedive to $56 billion. During the same period total farm expenses went from $306 billion to $376 billion. (And that's high, considering the good old days of the year 2000, when farm expenses were a mere $200 billion.) Not surprisingly, total farm debt during the same period from 2011 to 2016 jumped by nearly $80 billion.[5]

It's clear that Old MacDonald is getting squeezed. But flip through an issue of *Successful Farming* magazine and something else is obvious. While producers may be losing money and their farms, somebody is making money at farming. A lot of money.

Ray explains that for the farmer, the problem is on both the profit side of the balance sheet and the loss side. The operating overhead of a large farm is nothing to sneeze at. "A majority of their income is spent on inputs. Fertilizer, herbicides, pesticides. Chemical petroleum-based inputs are probably one of the most expensive things in modern agriculture, plus maintaining the infrastructure. Average producers have millions tied up in their operations. The risk is tremendous." Like

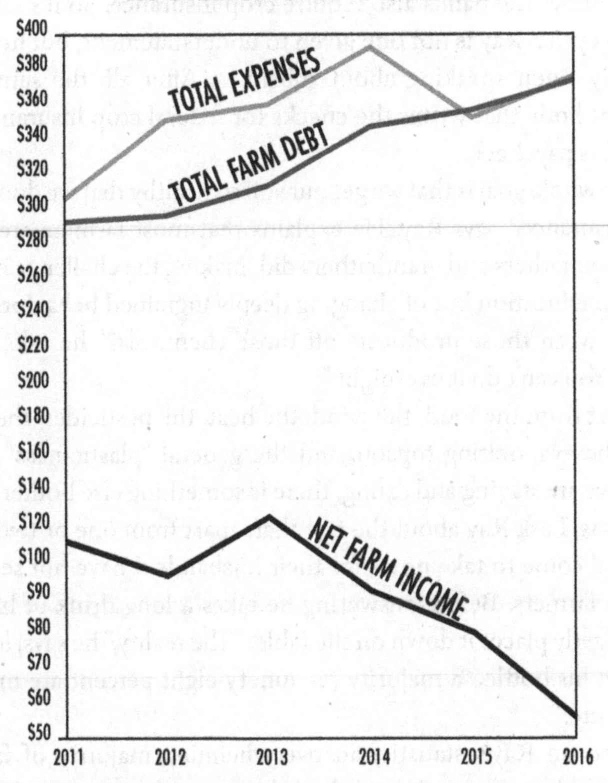

The end result of the Nazi chemical experiment that has become our modern industrial agriculture is that farm income is plummeting while farm expenses and debt are skyrocketing.

a bad night in Vegas, the only way out for many farmers is to double down by buying the latest machines, the most high-tech seed, and more chemicals—all the while digging deeper into debt.

I ask Ray why he thinks modern "precision agriculture" isn't helping farmers keep their farms. "They keep throwing in a lot of inputs, but the symptoms are more runoff, more disease, more pathogens; their crops can't handle the drought," he says. "But they know deep inside 'Well, if I have a crop failure, I can still depend on crop insurance and all my subsidies, and I still survive.' We also have to remember the banks also require crop insurance. So it's kind of a vicious cycle." Ray is not one given to understatement, but he treads carefully when speaking about subsidies. After all, the same government body that writes the checks for federal crop insurance also writes his paycheck.

"My whole goal is that we get our soil so healthy that we don't need crop insurance," says Ray. He explains that most farmers are doing what their fathers and grandfathers did, making the challenge not just one of reeducation but of changing deeply ingrained behavior. "How do you wean those producers off those chemicals?" he asks. "Very slowly. You can't do it overnight."

Apart from the food, the wind, the heat, the pesticides, the fertilizers, the evaporating topsoil, and the general "plastic-ness" of the places we are staying and eating, there is something else bothering me in Kansas. I ask Ray about the fact that, apart from one or two wives who had come to take notes for their husbands, I have not seen any women farmers. Before answering he takes a long drink of his beer and gingerly places it down on the table. "The reality," he says, looking up from his bottle, "a majority . . . ninety-eight percent are males in agriculture."

Based on Ray's statistic, an overwhelming majority of farmers (most of whom are male and don't have wombs to carry children) are the ones who spray extremely toxic chemicals (that require Tyvek suits and respirators) onto America's food. A woman, whose body and whose children are far more susceptible to the effects of chemical toxins, might at least think twice before donning protective gear in order to apply some type of -cide onto the very fruit, vegetable, or grain she will feed her children.

"When the ladies do come, they pick it up first because they're nurturers," Ray admits. Of course, that's a big "when." According to the USDA, women run a whopping 14 percent of the nation's farms, up from just 5 percent in the 1980s. Then there's the Women, Food, and Ag Network (WFAN), which "links and empowers women to build food systems." But as you might guess, WFAN is based on sustainable practices and its website is focused on farming that promotes the health of the people and the planet.

Ray, other producers, and the USDA itself confirm that women-run farms are slowly cropping up *along the coasts*. But when it comes to the bulk of US industrial agriculture (aka the farming belts), it's the middle of the country that counts. And when it comes to female farmers, that vast middle might as well be a desert.

I prod Ray, asking what other uncomfortable realities he sees on American farms. He takes a deep breath. "Oh, you want reality?" He's about to let loose. "Okay. Another reality, a majority of our acres is corn and soybean. Another reality, a majority of the land is degraded. Another reality, we have major water-quality issues. Another reality, we have massive desertification going on in our rangeland. Another reality is we have a hard time getting young producers coming into agriculture 'cause it's very costly. Another reality, our farms are too big. We farm way too many acres." Finally, Ray takes a breath.

All those "realities" don't exactly paint a positive picture for the future of farming. Ray insists that he is hopeful, there is light at the end of the tunnel, that producers are waking up every day and helping one another by building networks of support, et cetera, et cetera. But it is getting late and this part of his shiny brochure spiel is starting to sound as hollow as one of the empty beer bottles that the waitress has whisked away. I want the "real reality"—the world according to Ray. I ask my last question: "Ray, tell me honestly. Is it getting better or worse?"

"I have flown millions of miles. I have been in every state. I think it's getting worse," he admits quietly. "You're gonna have to have a temporary collapse. Not a total collapse. I think the system's gonna have to fall flat on its face. I really believe humans don't learn something until they go through a trial, until it gets so bad something has to happen." He lets that sink in for a moment. "This is a moral issue. This is a spiritual issue, and we're trying to fix it with pure science. It's not going to happen."

After five thousand miles of road travel listening to soil, desert, food, and agricultural experts across America, I am beginning to question everything from what I put into my mouth to the trajectory of our country and species. Like many of the other people I meet and interview who have dedicated their lives to studying soil, Ray seems to flip-flop between his ever-enthusiastic "can-do, let's help the microbes make the world a better place" attitude and his secretive premonition of a possible "soilpocalypse."

This is not his or any soilophile's fault. Blame rests with the human brain, which does not come prewired to deal in the magnitudes that one encounters when one begins to learn about soil. The "hundreds of millions" of acres of agricultural land in the United States, the "billions" of organisms in a teaspoon of soil, the "trillions" of underground life-forms, the "gigatons" of carbon dioxide that can be sequestered. It's almost as if everything that has to do with soil, from the destruction of soil itself to the possibility to build it back up and alter the planet's ecosystems, happens on both such an extreme micro and macro level that it hurts to try to comprehend it all.

The modern American agricultural system of production is indeed incomprehensibly big and at times wretched, but its existence has spanned a time shorter than that of the telephone, which only recently became radically more useful. While the soil as an ecosystem is complex, the methods Ray teaches are simple: cover the soil with a living plant at all times, don't till, use sprays extremely sparingly if at all, get ruminants and their manure onto the land, rotate crops, and always be a student of your soil's health.

Because of the simplicity of these tenets, the relative ease of implementation, and the material benefits of more food and more money per acre, the possibility for successful transformation of our soils through regenerative agriculture may be just as great as the possibility of failure. Perhaps that is why, despite the dire circumstances of our current farms and our soils, it's the optimistic side of people like Ray that wins the internal struggle between anguishing over the "soilpocalypse" and working toward a "soiltopia."

It's time for Ray to get the check, trudge through the wind, and cram in a few hours of sleep. Tomorrow is a new day. There will be a new group of farmers, another lecture, and at least one possible convert.

On his way to his room, Ray turns back toward me. "One more thing," he says. "Promise me you'll go see Gabe Brown in Bismarck."

"That farmer you keep talking about in North Dakota?" I ask him, wondering if North Dakota can possibly be any worse than Kansas.

CHAPTER 8

BISMARCK OR BUST

When one thinks of sunny summer tourist destinations Bismarck, North Dakota, does not likely come to mind. In fact, aside from a relatively recent shale oil and natural gas boom (which has since slowed due to a slump in oil prices), North Dakota generally sees few visitors.

The only states, in fact, that have fewer inhabitants than North Dakota are Vermont and Wyoming (the former being substantially smaller in area). Known for its big skies, buffalo, and brutally cold winters, this is the state that inspired the movie *Fargo*—a homely place where tough, pregnant cops chase criminals who put frozen corpses into wood chippers. It's a beautiful if somewhat harsh environment.

In many ways Gabe Brown is an apt expression of this place. His old T-shirt, tattered baseball cap, reddish round face, and full frame would make him indistinguishable in a lineup of "average American farmers." But spend ten minutes listening to his radio announcer-like voice and his nearly encyclopedic knowledge of food, agriculture, government, genetic engineering, economics, chickens, cows, corn, soil, carbon, or one of hundreds of other farm-related topics and you may find that Gabe is anything but average.

Perhaps that's because Gabe didn't have to follow rules of farming handed down to him from a long line of farmers. A self-proclaimed "city boy," he was born and raised in Bismarck. He became interested in agriculture in high school after taking a vocational education course in farming. He started working here and there on farms. It was in college where he met and fell in love with his wife. Lucky for Gabe, her parents happened to own a farm. (In his telling of the story, it's unclear whether or not he fell in love with the girl or the farm first, but I give the man the benefit of the doubt.)

Like most people I meet on this journey, Gabe gets up early. Before arriving I receive exactly two pieces of information from him through e-mail, his address and the words "Be here at seven A.M. I don't work bankers' hours."

Brown's Ranch is located about a half-hour drive from Bismarck, much of which is on dirt roads. Like most medium-scale farms, Gabe's house and farm buildings are clustered together. His home is a simple single-story brick building that looks identical to so many other farmers' homes. Surrounding it are a few barns and several hundred yards away are some small elevated grain bins. I only see one tractor and very little in the way of "big toys," the machines common to most medium- and large-scale farming operations in the United States.

No sooner do I step out of the vehicle onto the misty morning grass than Gabe rolls up on a four-wheeler. "Well," he says, surveying me with a blank expression, "you want to do your interview now? I'm ready when you are." Small talk is for folks with nothing to do. And Gabe has a lot going on.

I sit across from Gabe in a weathered metal barn that houses an old delivery truck, a tractor, and a few hundred bags of seed. Without much in the way of prompting, Gabe unfolds his life story in his unshakable North Dakota cadence.

In the beginning, things looked positive for young Gabe. His was a farm at the top of the Spring Wheat Belt and as such, his in-laws had farmed spring wheat, barley, and oats. With over six thousand acres under his plow, a new wife, and a new lease on life, the sun was just starting to rise on his little slice of agrarian heaven.

That was right around the time that Gabe decided to throw away all conventional farming wisdom and take a different path.

"For the first two years, I was farming conventionally—heavy, heavy tillage, high use of synthetics. Then in 1993, a good friend said,

'Gabe, in your dry environment, you need to go no-till in order to save moisture and time.' " North Dakota receives an average of seventeen inches of precipitation a year, making it one of the driest states in the country. Gabe continues, "Being a beginning farmer, I couldn't afford to just go out and buy no-till equipment, so I sold all my tillage equipment, and we've been one hundred percent no-till ever since."

Going no-till was definitely bucking the system. But it was what came next that put Gabe's mettle to the test. "I also started to diversify the crop rotation. I started adding things like peas and alfalfa and then corn. Well, 1995 came along, and the day before I was going to start combining I lost one hundred percent of our crop to a hailstorm. We were totally wiped out. No income from our crops. So that was tough for a young family starting out. Then 1996 came along, and we lost one hundred percent of our crop to hail again. Well, I had to start growing different crops just to produce feed for the livestock, 'cause the hailstorm took out all our feed." At this point, a sane person might have quit.

"Nineteen ninety-seven came along, and there was a major drought in this area. Nobody combined an acre. So I had three years of crop failure. Well, times were really getting tough financially, and the banker didn't want to loan me any money to buy inputs, so I really had to start focusing on how could I make this farm and ranch get these soils to produce a crop without all these inputs. So I started to be a studier of soil ecosystems, and I actually went back and read Thomas Jefferson's old journals, 'cause I was trying to figure out how they did this previous to the use of all these synthetics."

At this point, you'd think that God, nature, something, would have given Gabe a break. "To make a long story longer," Gabe says, "1998 came along, and we lost eighty percent of our crop to hail. So we were four years with basically no crop income, and my wife will tell you it was hell to live through. We both took off-farm jobs plus tried to run an operation of this size." He takes a deep breath. While his neighbors had all been whacked with one, two, or three failures, his was the only farm in the area that got hit all four times.

"The neighbors were all taking bets and waiting for my land. But I wasn't going to fail. Now I'll tell you, my wife was really questioning her pick of husbands at about that time, but fortunately she stuck with me."

Since his early trials and tribulations, Gabe Brown has weathered two more decades of farming and ranching. Because his ranch was not tilling the soil or using synthetics, some surprising things began to happen. "Over a period of time we noticed there was a difference to our soil," Gabe says. "It was softer to walk on. There was more life in the soil. I often tell people I could never go fishing the first ten years I was on this operation 'cause you'd never find an earthworm. Well, now you can't hardly dig up any soil without finding a handful of earthworms."

THE CHALLENGE

Gabe tells me, "I'm all about living things and taking part in living ecosystems. And to me that's what my operation's about—a living ecosystem."

Merriam-Webster defines *ecosystem* as "the complex of a community of organisms and its environment functioning as an ecological unit." Of course, this is a bit self-reflexive unless one already knows the definition of *ecology*. Derived from the Greek "oikos," meaning "house," ecology refers to the biological science of studying organisms and their relationship to their environment.

The truth is, we don't have a very good definition of *ecosystem* or *ecology*.

Biologists tell us that in nature, in places where man has had the least impact, the interactions between species and their environments contain nearly infinite layers of complexity.

One thing that's clear even in the early-morning mist that hangs around Brown's Ranch is that it's loud. Not concert loud, but loud enough so that my sound recorder is battling to pick up his voice over the crickets, cicadas, and goodness knows what kind of little creepy-crawlies that are making such a racket. Walk anywhere on

Gabe's farm and a wave of little jumping creatures scurry in every direction from your footfalls. There is most definitely a system of life here.

Modern agriculture seeks to remove as many of the complex interactions of nature as possible. In fact, the simpler the interaction (sunlight, water, chemicals, field corn) the easier it is to financialize, control, and predict the outcome. Gabe says, "Look at what's happening in production agriculture today. You walk into a supermarket and seventy percent of the things in that market have either corn or soybeans in them. That's just mind blowing. Look at what we're really doing in agriculture—we're producing monocultures of corn and soybeans."

The problem is, the more interactions you remove, the more you kill the elements that support the ecosystem. If we think about the Greek idea of ecology as a "house," one can only remove a certain number of walls before the ceiling crumbles inward.

Farming being hard work, I ask Gabe what the most difficult part of his job is. "The hardest part of my job is watching my neighbors," he says. "Approximately three weeks ago I stood here for a full day and watched two airplanes drop pesticides over my neighbors' fields. I had to feel that pesticide as it drifted onto my property. I was literally sick to my stomach because it is destroying everything I worked for by providing the home and habitat for all these insects, pollinators, predators, wildlife, everything. And it's being destroyed by a producer who thinks they are doing the right thing to produce food."

With all the farm tours and talks Gabe does, it seems like his methods would have caught on more, I say. Why are his neighbors still doing things the conventional way? "It's drilled into them every day through periodicals they read, through stories on the TV and radio, through the farm extension programs, through the government farm program. It's drilled into them that this is the way you produce; 'we've got to feed the world,' " he says with a hint of sarcasm.

"Feeding the world is going to be extremely difficult if we stay in the current production model. It is not going to be difficult if we go to a regenerative type model," he adds.

"Look at how natural ecosystems function. You have a tremendous amount of diversity. We raise corn, spring wheat, winter wheat, barley, oats, alfalfa, peas, triticale, hairy vetch, and clover. We're all about seeing how much life we can have on our operation.

"We run three hundred fifty cow-calf pairs and four hundred to eight hundred stockers, and they're all grass-finished, so we're grass-finishing beef," says Gabe. "Cow-calf pairs" are mother cows with their offspring, while "stockers" are animals designated for slaughter. Gabe continues, "We've got a flock of sheep. We grass-finish lambs. We've got pastured hogs. There's a thousand laying hens that are free range out on the pasture. We run broilers. We have an orchard. We started a bunch of nut trees this year. We produce honey off our own operation." Gabe says he's forgetting more than half of what he and his son grow because he doesn't have to manage everything. Instead nature does most of the work.

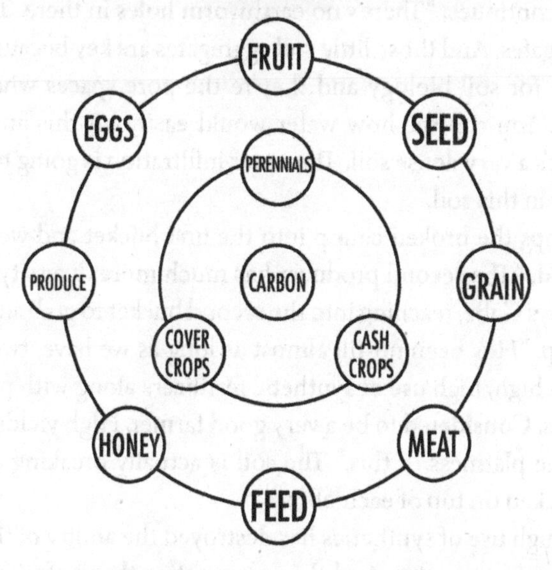

Whereas most farms today have only one income source, Brown's Ranch makes money from a diverse array of end products.

A POLYCULTURE OF POLYCROPS

We're both getting antsy sitting in the barn. Gabe gets me on his Polaris four-wheeler and drives a little way away from the buildings and down toward a large stand of crops.

We get off the Polaris, and Gabe hauls three white five-gallon buckets onto the ground. He's obviously prepared a little presentation. It turns out he was up at the crack of dawn driving around, stealing dirt from his neighbors. "I pulled three different soil samples from three different operations, all in near proximity to where we're at today," he says. He wants to show me a side-by-side comparison.

He bends down and takes a two-handful scoop of soil from the first bucket. "So the first producer has very little crop diversity. Only grows spring wheat and flax. Has been no-till for about, oh, fifteen years. Does use pesticides, herbicides, synthetic fertilizer, anhydrous ammonia. It's pretty easy to tell by the soils." He breaks the soil clump open. It's like a rock—light brown and very hard.

Gabe continues, "There's no earthworm holes in there. There's no soil aggregates. And those little soil aggregates are key because they're the home for soil biology and they're the pore spaces where water infiltrates. You can see how water would easily hit this and run off because it's a very dense soil. The water infiltration is going to be very, very poor in this soil."

He plops the broken clump into the first bucket and we move to the second. "The second producer has much more diversity, and he's no-till," says Gabe, reaching into the second bucket to grab another big soil clump. "He's been no-till almost as long as we have, twenty-plus years. But high, high use of synthetic fertilizers along with pesticides, fungicides. Considered to be a very good farmer. High yields. But just look at the platiness of this." The soil is actually breaking apart like plates stacked on top of each other.

"The high use of synthetics has destroyed the ability of this soil to produce soil aggregates. And then also notice the roots. They move horizontally in the soil profile. They're not able to move down into the soil profile," says Gabe as he points out how the roots actually travel sideways along the soil "plates."

I now have my own clump of farmer number two's soil and am inspecting it with some fascination. "So more or less, this operation is surviving just on inputs," says Gabe. "I would say this epitomizes the current production model. That's what our current farm program and policy has done for the soils. Also notice that there are no earthworms. It's going to be very difficult for water to infiltrate in there. You're not going to be able to move carbon through photosynthesis deep into the soil profile if your roots can't penetrate. It's a tough environment."

We drop farmer number two's platy soil into the respective bucket and move to bucket number three.

"This producer here is an organic producer," explains Gabe as we pick up soil clumps from the bucket. "A lot of diversity. Obviously no synthetic fertilizers, pesticides, fungicides, but you have no soil aggregation because of the tillage." The soil breaks apart in my hands and turns to dust. Letting the soil sift through his fingers, Gabe says, "It's just . . . it's a powder of a soil that's not able to hold on to water. What moisture he gets is going to evaporate away, because there is no armor on the soil surface and there's no soil aggregates, no organic matter to hold that moisture. Pretty typical of most organic operations."

I understand why Ray Archuleta wanted me to come meet Gabe. The Gabe Brown "soil tour" is reminiscent of Ray's. Both Gabe and Ray believe that large-scale organic producers often inadvertently damage their soils. Neither man is antiorganic, but seeing Middle America through their eyes, it doesn't matter how you farm: heavy tillage takes an immense toll on the soil.

Gabe wipes the remainder of the dirt on his old jeans and says, "Now we'll walk over and look at some of my soils." He grabs a shovel from the back of the little four-wheeler and heads over toward a five-foot-high wall of plants.

"What we have here is my idea of a larger-scale garden," he says. "There are about four acres here, and they're planted with sweet corn, zucchini, watermelon, cantaloupe, squash, peas, beans, but they're all planted together. So instead of a monoculture, what we have here is these roots all intertwined. Mycorrhizal fungi, which is critical to moving nutrients throughout the profile, is transferring nutrients from one of these plant species to the other. So they're all working in symbiosis. The other thing you can see just by looking is the large amount of insects. See the butterflies. You can see a lot of ladybugs. You hear the crickets. I mean, it's alive."

He parts the large stalks of plants and heads into the dense growth. I follow him in and am instantly covered with water as the dew from the plants wipes off onto my arms and legs. "Now, you can see how easily I can stick a spade into the soil." He slides the shovel down into the earth and breaks open the surface. The rich, earthy-smelling soil clumps nicely together. As if on cue, a worm wiggles out. "There's some earthworms right there. But look at this soil; notice the dark color? That's indicative of higher carbon. More carbon is moving through that liquid carbon pathway in these soils," says Gabe.

He picks up his clump of soil to show me. "You notice the aggregation?" he asks, showing me the little clusters of soil clumping onto the roots of the plants. "Those are all soil aggregates or soil particles, and a soil aggregate will live approximately four weeks, and then you've got to form new ones. The only way to form new ones is with biology and glomalin, which is a glue secreted by mycorrhizal fungi.

"It doesn't take rocket science to see these roots are going to have

Gabe in his polycropped garden. (*Simon Balderas*)

no problem moving down into the soil profile," Gabe continues, showing the spindly roots hanging down from the soil clump. "And even though we haven't had much moisture in two months, you can still feel that this soil is damp. You can also see all the earthworm holes in here. It's full of life, and that's what we want to see in a healthy soil."

Gabe says that when he and his wife took over their ranch in 1991, their soils had organic matter of 1.7 to 1.9 percent, which is roughly equivalent to how many inches per hour of rainwater that the soil can infiltrate. He estimates that the soils of the area had once been 7 to 8 percent before the Europeans arrived, with corresponding infiltration rates of seven to eight inches per hour. Now his soils are around 6 percent organic matter.

"For every one percent organic matter a soil will hold twenty to twenty-five thousand gallons of water." Brown's Ranch could hold fifty thousand gallons of water per acre when its soils were at 2 percent organic matter. "Now I'm near six percent, and I'm holding a hundred fifty thousand gallons of water in that soil profile," Gabe explains.

Gabe's firsthand experience with the soil-water connection has left him with a strong opinion about what is happening in California. "I was there in February, and they had thirty-two inches of rain on this one operation. They're in a 'drought.' I told him you're getting twice the rain I get in a normal year, and I'm producing crops," he says. "They don't have the healthy soil where they can infiltrate water, store it, and have it available for the plants. They're creating their own drought, plain and simple. Agriculture is the answer to the water cycle because you create your own microclimate."

Gabe breaks his lecture and crouches back down to the ground. "The other thing is look, there—armor on the soil surface," he says, running his hand over the thick layer of thatched vegetation that covers the ground in between the tall stalks of vegetable plants. "This soil is never bare. So if a raindrop falls, hits this, that's tremendous force. We're not going to get the erosion. We're also not going to get evaporation even in these high temperatures."

Gabe stands up and motions to the cornucopia of plants surrounding us. "This is a polyculture, and to me it's just beautiful."

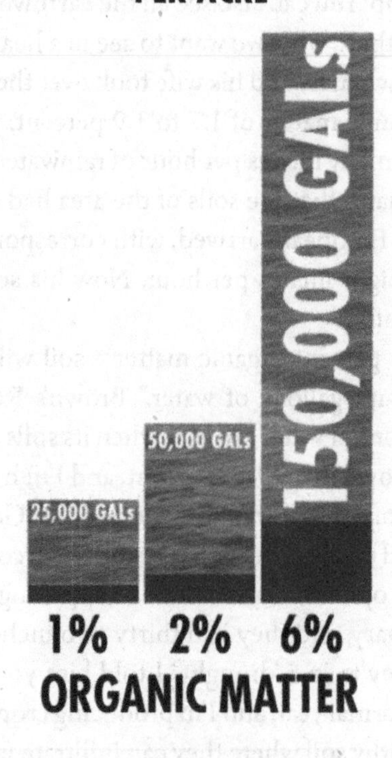

As Gabe Brown has increased the organic matter content of his soils, those same soils hold more water.

As opposed to a monoculture, in which one crop is grown, in a polyculture there may be two, three, four, tens, or even hundreds of species that are intentionally planted and grown together as "companion plants."

I ask him how he gets all these vegetables out of his "garden" without machines. "Large-scale production ag will say, 'you can't do this large scale.' Well, I can do it large enough scale to feed my customers. We walk through here and harvest it by hand." When I ask him if such labor-intensive work is practical, he asks, "Isn't there an unemployment problem in this country?"

He says this particular garden has another purpose. "We enjoy giving a lot of this away. So we will let needy families come. Do you know how much the kids love to come? It's like Christmas to them. And those kids are learning about where their food comes from. Now does it take manual labor? Yes, but not as much as you'd think."

Even though the garden is a minor part of his enterprise, it does make money. Explains Gabe: "There are four acres of this mix right here. So far we've harvested most of the peas. The zucchini is about done for the year. The beans are getting down to the tail end. We're harvesting the sweet corn now. We still have the watermelon, the cantaloupe, and the squash. We've already made over a thousand dollars an acre with this. Now most grain farmers hope to make a hundred dollars an acre. They're ecstatic about that. We're already over a thousand dollars an acre and we haven't even harvested half of what we have here."

Gabe says there is yet another layer to his garden's profitability. After the fruits and vegetables are harvested from his garden, he'll move the cows in. They will forage for edible plant bits and trample down much of what is standing. The cows will be followed by his chickens, which will eat things on an even more granular level.

Gabe reaches over and grabs a tall stalk of corn. He rips off an ear, pulls back the husk, and throws it over to me. "Go ahead, take a bite," he says. I chomp down on the raw corn and flavor explodes in my mouth. It tastes different from any corn I've ever had. When I ask Gabe about it, he says it's just standard non-GMO "sweet corn." "I'm not saying you're not going to find an insect on here, but that's just part of the cycle," he says, handing me another ear of corn. "This one will be good too. You don't even need to cook it." Compared to my Starbucks breakfast, which tasted like cardboard, this corn is sweet and filling. It tastes like nature.

SUN FUEL

Gabe is driving me between his fields. Like most medium-scale operations in the United States, his five thousand or so acres are split into parcels known as 160-acre "quarters" and forty-acre "quarter-quarters" (160 acres being a quarter of 640 acres, which is 1 square mile, the once idealized size for a township). Having land parcels divided by roads, other farms, subdivisions, and industrial areas makes farming infinitely more complicated. Gabe says he has friends whose farm fields are tens of miles, even hundreds of miles, apart.

He takes his truck down a dirt road a few miles from his homestead and pulls off onto the side of the road. I follow him into a field of what appears to be weeds.

Gabe explains that his goal on much of his land is to mimic the type of plants that a native prairie would have had prior to European settlement. He starts pointing to all sorts of plants.

"Here, I have hairy vetch. Hairy vetch is a biannual legume. It'll be twenty-five to thirty percent crude protein. It'll fix a tremendous amount of nitrogen. That nitrogen, then, through mycorrhizal fungi, feeds the grasses, the sorghum-sudan and the millets. Here is soybean—another legume. That's gonna fix nitrogen. This is red clover, another legume that fixes nitrogen. There're sunflowers in here. They have a deep taproot, so they break up compaction layers. There's a little bit of flax in here. Flax really forms a symbiotic relationship with mycorrhizal fungi. So it helps proliferate the mycorrhizal fungi in the soil. There's buckwheat—a flowering species that scavenges phosphorus. It is also an excellent habitat for the pollinator species. The sorghum-sudangrass that is in here has a very fibrous root system, and it'll really increase organic matter—plus it's high in nutrient quality for the cattle."

Gabe is excited about each of these species in what he calls this "mix," meaning the mixture of seeds. He says he typically plants a mix of nineteen different species that more or less mimic what likely once grew in the "natural" prairies of this region. Like a certain instrument

GABE BROWN'S MULTISPECIES CROPS

PLANT	FUNCTION
HAIRY VETCH	BIANNNUAL LEGUME - 25 TO 30% CRUDE PROTEIN - FIXES A TREMENDOUS AMOUNT OF NITROGEN - THAT NITROGEN, THROUGH MYCORRHIZAL FUNGI, FEEDS THE GRASSES, THE SORGHUM-SUDAN & THE MILLETS.
SOYBEAN	LEGUME - FIXES NITROGEN.
RED CLOVER	LEGUME - FIXES NITROGEN.
SUNFLOWERS	HAVE A DEEP TAPROOT, WHICH BREAKS UP COMPACTION LAYERS.
FLAX	FORMS A SYMBIOTIC RELATIONSHIP WITH MYCORRHIZAL FUNGI - IT HELPS PROLIFERATE THE MYCORRHIZAL FUNGI IN THE SOIL.
BUCKWHEAT	A FLOWERING SPECIES THAT SCAVENGES PHOSPHORUS - ALSO IS AN EXCELLENT HABITAT FOR THE POLINATOR.
SORGHUM-SUNDANGRASS	HAS A VERY FIBROUS ROOT SYSTEM - INCREASES ORGANIC MATTER - HIGH NUTRIENT QUALITY FOR CATTLE.

Brown's Ranch uses cover crops to feed carbon and nitrogen into the soil, eliminating the need for synthetic fertilizer. Cows then eat the cover crops.

in a choir, each species in his mix has a specific role or function. As a result of the mix of plants, the movement of his livestock through the fields, and the various other crops he plants in alternating years, the actual count of plant varieties is much higher. Says Gabe, "This is what I like to see because this is as close to a true native ecosystem as we can get where we're located."

We walk deeper into the dense vegetation so I can get a good photo of Gabe in the morning light and I ask him how he makes a profit on a cover crop like this. He says it costs him twenty-eight dollars in seed to plant an acre. The economic upside comes from his animals, which eat the food he grows in that acre. By simply "growing feed," Gabe reduces the cost of his inputs for his animals to pennies on the dollar of what other ranchers pay. That means less cost, more profit, and better soil.

He explains it like this: "Every farmer and rancher, their whole business centers around taking sunlight and converting it into some product. So every leaf out there is a solar collector. Well, the more and different leaf sizes and shapes you have out there, the more solar collectors you have, the more energy you're gonna be able to draw into your system, and you're gonna be able to convert that to carbon. The more carbon you have, the higher your potential profitability."

We trudge through the brush and get back into the pickup. While Gabe drives, he continues his Farming in Nature's Image 101 lecture. He's still talking about the importance of planting multiple species together. "Where, in nature, do you find a monoculture? You don't. Usually only where man put it. So nature abhors a monoculture," he says.

THE CORN GAME

The following morning, we head out in Gabe's tractor so he can show me his "no-till drill." The device is built by John Deere and looks exactly like a disc harrow or tilling machine with one big exception. It has fine blades that delicately "slice" into the soil like razors, then seeds come down, go into the slits in the soil, and the soil is firmed back into place with big rubber wheels that "seal" up the slits. Gabe says that most of the farms in the central part of North Dakota are no-till, but unlike Gabe they have not made the shift from monocropping to polycropping.

Gabe says that farmers are extremely risk-adverse and that makes them terrified of change. "I argue that if you build a healthy farm ecosystem, it's going to take the risk out of it, because you're going to be resilient."

Gabe says that most growers in the corn and soybean belt are completely dependent on income from just one or two crops. He contrasts this to his polycropping model. "I also grow wheat and barley and oats and vetch and triticale and alfalfa and grass-finished beef, lamb, pork, honey, vegetables, and all these other products. I could care less what the price of corn or beans are 'cause they're such a small fraction of my operation. I built resiliency into my ecosystem," he explains.

He says that while it costs between $3.50 and $5.00 to grow a bushel of corn, depending on where you are in the country, it only costs him around $1.40 a bushel. Even when the price of corn plummeted to $1.77 a bushel, Gabe still made a profit.

When I ask him what he thinks is the biggest impediment to farmers making the switch to a more diverse model of farming, his answer is lightning quick. "The federal farm program we have today is the most detrimental thing there is to regenerative agriculture," he declares. "It dictates that producers—if they want that safety net through crop insurance—can only produce monocultures. They can't do what I'm doing."

During the cold North Dakota winter months, when farming slows down, Gabe hits the road, speaking to farmers' groups all across the country. He describes one of his lectures. "I was speaking to a group of corn and soybean producers from Missouri this past winter and one guy stood up and said, 'Gabe, why would I want to plant anything else than corn and beans when I'm guaranteed my income? As long as I keep my expenses below a certain point I'm guaranteed a profit before I even go in the field in the spring.'"

Gabe's reddish face is now almost the color of a tomato and his northern voice is at full radio volume. "How do you answer that question?! I mean where else in business are you guaranteed to make X amount of dollars every year before you even go in the field?"

Gabe is on a roll now, and I don't dare interrupt. "Now I'm going to be much more profitable than he is, even though he has that guarantee. And then I come back and ask them, 'How much money would you be making if the government wasn't subsidizing your operation?' They know the answer. They're not going to be making money, and that's all there is to it."

Gabe says even the promise of higher profits is not enough to entice most farmers off the government teat. "It bothers me," he says. "We're not guaranteeing 'Ma and Pa's hardware store' on Main Street—we're not guaranteeing them a certain income level. We're not paying part of their insurance, which is what's happening with the federal farm program." Gabe is of course referring to the federal crop insurance program that doles out money to farmers who grow "acceptable" commodity crops using "acceptable" chemical sprays.

The commodity insurance game is what Gabe refers to as the "vicious cycle." He explains it like this: "Last year, everybody was planting corn and soybeans because they could lock in the most revenue insurance with corn or soybeans." And what happens when the market gets flooded with too much of something? Predictably the price goes down. And that's exactly what happened with corn and soybeans during that cycle. The price plummeted. So the Risk Management Agency (RMA) adjusted insurance prices for the following year.[1]

Gabe says, "This year, sunflowers locked in a little more profit, so now there's a lotta sunflowers being produced. That'll guarantee the sunflower price will go down next year. It's a vicious cycle. Farmers are their own worst enemies."

Gabe is a staunch free-market advocate. Having survived years of no crops because of drought and hail, he believes farmers can thrive without the government. "I dropped off all those programs. I no longer take part in federal crop insurance," he says with steadfast conviction. "I tell producers I don't want to be on welfare anymore. And that's what I look at it as—welfare."

As a regenerative farmer, Gabe Brown has found a more important guide of what to plant than the US government's crop insurance program. "Those type of programs limit my ability to adjust to what nature is telling me," he explains. "If we have a very open winter, which is what we call it here when we don't get much snow, that's going to limit our moisture. I can adjust what crops I'm going to plant. The producers who are only growing corn and beans—they're locked into that. I'm not. I've got the best of all worlds because I can fluctuate and adjust accordingly.

"Now they're going to say we're doing that to ensure a safe and healthy food system, ensure cheap foods, so to speak. Well, I don't want cheap food. I want nutrient-dense food and there's a big difference." Gabe claims that even with crop insurance and the newest gadgets and the best genetically modified seed, other farmers cannot compete with the amount of per acre food he produces. He believes "cheap food" is part of the "feed the world" campaign wherein farmers repeat the slogans given to them by chemical companies and the government only to find they are doing the exact opposite.

"Look at the facts," he says. "The average corn production in

Burleigh County, North Dakota, is just under one hundred bushels per acre. My proven yield on our farm is one hundred twenty-seven so I'm producing twenty-five percent more than the average in this area and I'm not using GMOs and I'm not using all the fertilizer, pesticides, fungicides, et cetera. So why do we need it? Why do they need the drought-tolerant gene in corn? Because they've destroyed the soil and they now have continuous drought. It's ridiculous."

Herein lies yet another layer of the vicious cycle of modern farming. Farmers with bad soils must use more and more technology to try to emulate what nature already does. Meanwhile the demand on them from the futures markets, the chemical and seed companies, the government and farming culture in general, is to produce ever more per acre. Their cost of inputs continues to rise. Their soil degrades. Their per acre profit continues to decline, often into the negative. Federal insurance must cover the difference. The following year, the cycle kicks into slightly higher gear and starts again. In this game of diminishing returns, we are literally paying farmers to go out of business. Like frogs slowly boiled in water, most farmers do not jump out until it's too late.

If insanity is doing the same thing over and over and expecting different results, then our food system qualifies as clinically insane. Perhaps this is why Gabe Brown concludes by saying, "I have no desire to produce commodities. I want to produce nutrition."

CHICKEN AND EGG

I am standing in a field surrounded by chickens several hundred yards from Gabe's house. There are no fences and no means of keeping the chickens from leaving.

Gabe's son, Paul, who is taller and more "millennial-optimistic" than his father, has been giving me a tour of cow dung beetles. Paul explains that the chickens follow about a week after the cattle have been through a field. The chickens eat dung beetles and fly larvae, thereby keeping the fly population down. "Anytime you can use the livestock to harmonize with each other, it's a win-win situation," says Paul. A chicken pecks its way across our path. "And it's free feed for our chickens," he says.

Farther up the field there are a couple of old horse trailers that serve as chicken roosts. Gabe drives up and parks next to them and we head over to meet him.

"These are what we call our 'eggmobiles,' " says Gabe proudly. "These will follow the cattle. These are built from old stock trailers that used to haul cattle or horses, and we can pick them up really cheap. We tear the floor out, put wire mesh in the floors, put roosts in, and hang the nest boxes. We have approximately $1,250 in each trailer in materials. So we can house about a hundred twenty-five to a hundred fifty laying hens in each. Right now, we have four of these on the operation, and between pullets [young hens] and laying hens, there's about a thousand of 'em out flying around."

Gabe opens the back door of the trailer, and I look inside while he and Paul begin to gather eggs. The chickens' nests are arranged in neat rows along the sides of the trailer. The mesh floor allows their poop to fall right on the ground. Gabe motions to a small door where the chickens are now exiting. "The door on the eggmobile is photosensitive, so once it gets toward dark, all the laying hens'll move inside to roost. That door'll close automatically. Then in the morning, when the sun comes up, the door'll rise, and out they'll go. That's to keep the predators out.

"Chickens are kind of our sanitizer," says Gabe as he uses the upturned front of his shirt to collect still more eggs. "We use them to sanitize the land, so to speak. So, after the cattle have moved through an area and grazed, then we bring in the chickens. In this case, it's the laying hens that we brought in. Those laying hens sanitize the soil, more or less, of any pest insects. This is what chickens were evolved to do: to scavenge things, and they live on insects.

"They also generate considerable income for us in that we sell pastured eggs from them," Gabe tells me as he carefully puts the mountain of eggs he gathered into his pickup. A chicken flaps into the truck and Gabe gently shoos it out. "Chickens lay according to day length, so during the winter, they tend to almost shut off laying, and we let them do that because it's the natural cycle."

He continues, "Otherwise, the eggs that are produced in conventional agriculture, those hens are all housed in little crates indoors and they don't get exercise, they don't get out, and they have to lay year-round. So most hens in an industrial model only last one to one and a half years. We have hens here that are seven to eight years old and still in production. So it's a healthy life for a chicken. Unless a predator comes along, but that's just part of the cycle."

Since Gabe's truck is stuffed with too many eggs, I head back up to the main buildings with Paul. The Browns have a building dedicated just for eggs with special refrigeration, storage, and cleaning equipment. Apparently it took some serious time and money to put together and was a necessary USDA requirement in order to sell eggs. I watch as Gabe and Paul put the eggs through a conveyor belt–type machine that cleans the eggs and passes them under an intense light to check for flaws. Then they carefully put each clean egg into crates. It is surprising how much personal care they take with each egg.

When the Brown men are done cleaning and crating their eggs, we head toward the house. Yesterday, Gabe instructed me to go buy a dozen "free-range" eggs from a local grocery store in Bismarck. I fetch my dozen eggs and Paul brings a box of Brown's Ranch eggs.

I find the Brown father and son on the porch of the simple beige brick home. The barbecue pit is open and an iron skillet has been placed there. It's time for the "Great North Dakota Egg-Off." Paul opens both boxes, randomly selects an egg from each, cracks each open, and plops them into the pan. The pale yellow yolk of the store-bought egg runs into the white. Meanwhile, the yolk of the Brown's Ranch egg is dark yellow and stands up like a bubble.

Like a sportscaster on *Monday Night Football*, Gabe begins commenting on the eggs. "There's a profound difference in the color, texture of the egg, and that's all due to what these hens eat. Our hens are out there eating bugs, crickets, grasshoppers—whatever they can find—along with some green plant material. Healthy. The other, even though it says 'cage-free, all-natural,' is, no doubt, in a building, no doubt has access to the outdoors, but it's being fed grain products, not bugs. And it's not able to do what a chicken should do."

I am upset my eggs are doing so poorly in this competition. After all, I bought them in good conscience and paid top dollar as a consumer wanting some of the best "free-range" eggs that a Bismarck grocery store has to offer. Paul is moving the pan side to side. My store-bought egg has scrambled itself. Theirs, meanwhile, is still intact. "The store-bought are typically more watery. That's what I've noticed," says Paul.

I tell them that no good competition is a one-off. We agree that the winning egg will be the best of three tries. When the third pair of eggs is put into the pan, I can see that my eggs have lost. Gabe likes to do a "crack test" to show me that their eggshells are harder than the store-bought eggs. Then he does a "poke test" to show me how many finger pokes it takes to get the yolk of the egg to break open. I feel ashamed as my eggs fail on all counts.

Gabe and Paul seem only mildly happy with their egg win. "We often get our eggs compared to organic eggs," Gabe tells me. "Just because a product is organic does not necessarily mean it's nutrient-dense." He says it is because the organic standard does not specify that the hens actually live outdoors, rather just that they must have "access" to the outdoors.

Just when I think the egg competition is over, Paul pulls out a small device that looks like a miniature telescope. Gabe explains that the thing is called a refractometer, costs around $100, and is typically used in the grape industry to see when grapes reach their peak sugar content on the vine and also in the home-brewing community to see when a mixture is ready. Technically speaking, a refractometer measures the "index of refraction" or rather, how light passes through a liquid.

In the food industries inside which the type of simple handheld refractometer the Browns have is used, a liquid is generally smeared onto the "reader" end of the optical device. The other end, the "viewing" end, is placed up to the eye. To see the "reading," the user simply aims the refractometer toward a light source (usually the sky). Inside the refractometer there is a scale that generally goes from zero at the

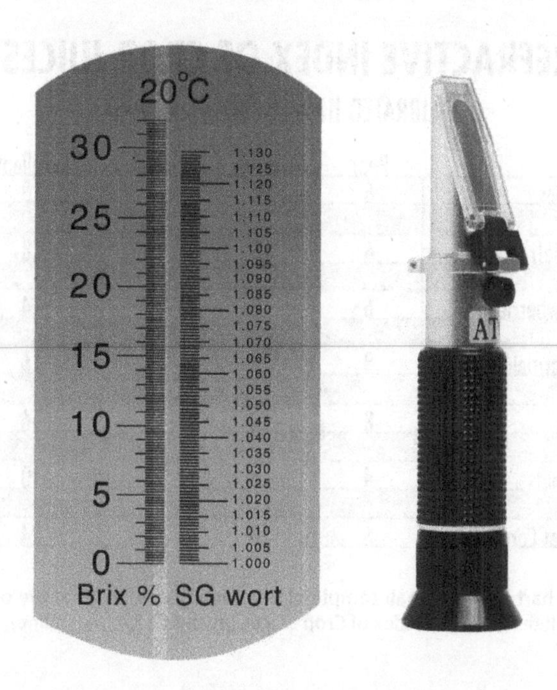

A simple handheld refractometer costs around $100 and uses the Brix scale.

bottom to anywhere from ten to ninety at the top. This is the Brix scale.

Technically written as "degrees Brix" or "°Bx," one degree Brix is equal to the density of one gram of sucrose dissolved in one hundred grams of solution. But sugar is just the standard by which Brix is calibrated. In reality, Brix is actually measuring the specific gravity, or rather density of a liquid. So it is really measuring the dissolved solids in the liquid. Brix therefore provides a loose measurement of the sugars, carbohydrates, proteins, vitamins, minerals, amino acids, and oils in a given food.

An agricultural engineer named Carey Reams did a series of experiments from the 1940s to the 1960s and created a Refractive Index of Crop Juices, which some ecologically minded farmers still use today

REFRACTIVE INDEX OF CROP JUICES

CALIBRATED IN % SUCROSE OR °BRIX

	Poor	Average	Good	Excellent
Apples	6	10	14	18
Bananas	8	10	12	14
Grapefruit	6	10	14	18
Oranges	6	10	16	20
Strawberries	6	8	12	14
Tomatoes	4	6	8	12
Watermelons	8	12	14	16
Carrots	4	6	12	18
Kale	8	10	12	16
Lettuce	4	6	8	10
Peanuts	4	6	8	10
Potatoes	3	5	7	8
Sweet Corn	6	10	18	24

This chart is just a small sample of the numerous foods that are on the complete Refractive Index of Crop Juices (available for free online).

to determine the nutrient density of the crops they are producing. The chart, which is available for free online through the Bionutrient Food Association, shows the Brix readings for Poor, Average, Good, and Excellent for everything from apples to sorghum to turnips.[2]

Gabe and I watch as Paul smears egg-white goo onto the refractometer, places it against his eye, and points it at the sky. "One of the things we've been trying to do in our operation is discover how we really know and understand that we're producing healthy food. Well, one of the simplest tools we have to measure this is the refractometer," explains Gabe.

Gabe says they test their own produce, their grains, and even the nutrient density of the cover crops for feed to ensure they move their livestock when the nutrient density of their feed is the highest, which is usually in the afternoon due to photosynthesis. They use it to ensure their hay crop is at its peak nutrient density before harvesting it and baling it for the winter. He has even gotten into the habit of testing the few off-farm foods his wife buys. "Now, obviously, I don't walk into a grocery store and test it right then and there. That will get me kicked out. But it's a simple tool to use to test how nutrient dense your food is," he says.

Paul gives me the refractometer. My store-bought egg produces a murky line on the display. The blue section at the top of the image blurs into the white section at the bottom and I can't really tell what the Brix percentage is. I wipe it clean and put some of the Browns' egg-white goo on there. When I look into the reader there's a razor-sharp line between the blue area at the top and the white at the bottom. While I have no reference for how this compares to the rest of the eggs on Earth (other than my runny store-bought eggs), the reading is high on the Brix scale.

YUPPIES AND HIPSTERS

The afternoon sun is hot, and Gabe is sitting on an old chair at the front of a barn that houses a menagerie of dust-encrusted machinery and tools.

"You know," he says, "it's often said farmers and ranchers are the only ones who buy at retail, sell at wholesale, and pay the freight both ways. I don't want to do that. I want to sell at retail and I want to capture as much profit as I can."

Gabe wants farmers to create their own markets, cut out the grain and futures traders and the supermarket chains, and sell directly to consumers. He says, "It's going to take the consumers demanding more nutrient-dense foods. Once the consumers start demanding that their food is produced in such a way, that will drive change in production agriculture."

He takes me over beside the egg barn and shows me their concessions trailer, which is being prepared by Paul and a female intern named Shalini to go into town to deliver orders. It looks like an ice cream trailer. Inside are large freezers filled with meat cuts. Outside it is adorned with a logo and the words NOURISHED BY NATURE, the brand he and Paul thought up for their Brown's Ranch products. Gabe picks through a few items in the freezers. "There's Nourished by Nature rib-eye steaks, Nourished by Nature beef, Nourished by Nature lamb, Nourished by Nature pastured pork, Nourished by Nature free-range eggs, and so forth.

"We can't keep up with demand right now, which really surprised me in Bismarck, North Dakota, that there would be that type of demand for local foods, but it goes to show you that this movement is growing. People want to know what they're eating and where it comes from," says Gabe.

It's almost time for Paul to hitch up the Nourished by Nature trailer and head into town. Paul explains that "a customer will place their order online, and a day before we're gonna deliver that order, we send them an e-mail with their exact total and location where we'll be for their order pickup."

Gabe says, "The question we get asked when we set up at a farmers' market is 'Where are you from?' That's always the first question ninety-nine times out of a hundred, 'cause they want to put a face with the food that they're gonna put in their children's mouths. They want to know 'Hey, if something's wrong, where do I find this guy?' Second question ninety-six times out of a hundred is GMOs—are you non-GMO? And here we are in Bismarck, North Dakota, where I wouldn't think that many people would be concerned about that. Third and fourth questions we get asked are antibiotics, synthetic hormones—are we using them in our meat products?"

I head to the park in Bismarck and find a large CSA (community-supported agriculture) trailer that is handing out its weekly boxes of veggies to members. Parked across from it is the unmistakable Nourished by Nature trailer. Paul and Shalini are hard at work doling out their products. Their customer base looks exactly the same as a random sampling of West Coast Whole Foods shoppers or farmers' market patrons—yuppies in after-work attire and hipsters in youthful flannel and denim. But the majority of the folks are just "normal" people. The reasons they cite for buying from the Browns are "health" and their "children's health."

Paul tells me that "getting to take a product from birth to harvest and then getting to connect to those people who are thriving from the products you provide is awesome. That's a big disconnect in the food system now: People raise food and then they dump it off and let someone else deal with it. They're missing out on the most rewarding part, and that's getting to talk with the consumers."

That's one part of the family business that Paul enjoys. He smiles and greets people by name, hands them their goods, and checks them off the list. By the end of their time at the park, they have pulled in a few thousand dollars. Not bad for a Wednesday evening in sleepy North Dakota.

ORGANIC-ISH

Gabe is almost practicing "organic" agriculture, but he does so without certification. It's my last morning on Brown's Ranch and I ask him why.

Gabe says, "My certification is by my consumers. I honestly think that the consumption of food and the act of selling food to a consumer is trust. That consumer trusts that the product I'm producing is as healthy as I can produce it.

"I can only think of a couple times I've been asked, 'Are you an organic producer?' and once I explain what I'm doing and why I'm not, we have never lost a sale because we're not certified organic," explains Gabe. "Organic, it's good; you're not using synthetic fertilizer, GMOs, fungicides, or pesticides. That's great. I'm all for that. But what does it say about nutrient-dense foods? Organic may or may not be nutrient dense. If we truly want a healthy society, we need to have nutrient-dense products. That's the standard I think we gotta reach."

As Gabe puts it, the big issue is not whether or not to be certified "organic," but rather whether or not to till the soil. "One of the things I wrestle with personally every day is do we till or do we spray an herbicide?" says Gabe. "I have a desire to be no-till organic. We can do it very small scale, a vegetable garden, et cetera. But to my knowledge, there's no large-scale operation in the world that's no-till organic and here's why: Nature wants the soil to be covered all the time. Well, weeds come. And why do weeds come? Because it's natural succession, nature's trying to put armor on the soil to protect it from blowing away, washing away, et cetera."

Gabe says that on a large scale it is very difficult to plan crop rotations so that the weeds are always eaten by animals. He has eliminated all pesticides and fertilizers but still finds that in order not to till his soil, he has to occasionally apply an herbicide. "We're down to about one pass every three to four years so I am eliminating it. But I haven't quite got there yet," he says. "I'm not going to give up. There's myself and many other producers around the world who are working to get there. There's many good researchers who are trying to get there. We just haven't figured out the right sequence of crop rotation and animal impact to totally eliminate it."

Gabe says using the herbicide is a difficult ongoing choice but one that he makes very carefully. "We try to use herbicides that have the least amount of impact on the biology of the environment. So we'll use those herbicides only when it's necessary," Gabe tells me.

"Are we doing damage? Absolutely. And it tears me up every time I have to apply one. But in my mind, the damage from the herbicide, the effect it's causing on my ecosystem, is much less damaging than a

TO TILL OR NOT TO TILL?

PROS	CONS
LEVELS LAND	OXIDIZES OR BURNS OFF CARBON FROM THE SOIL
ERADICATES WEEDS	RELEASES R. STRATEGIS BACTERIA, WHICH FURTHER EAT CARBON
BREAKS UP SOIL CLUMPS	BREAKS THE MYCHORRHIZAL FUNGI
MAKES CARBON AVAILABLE IMMEDIATELY	KILLS MICROBIAL LIFE
	DESTROYS "POROSITY" OR WATER-HOLDING CAPACITY OF SOIL

tillage trip across the field. 'Cause what happens with tillage is we're destroying the home for all the biology, we're destroying mycorrhizal fungi; we're going to inhibit water infiltration, and water is a huge issue. Because of us destroying all those things the soil is going to have less ability to move nutrients to those plants and that equates to lower nutrient density in the plants, which in turn equates to lower nutrient density in the product we're selling, and I'm not willing to do that."

Gabe says he could easily do no-till organic if his farming and ranching operation was smaller, but he prefers to perform agriculture at a scale common to North American farms. "Let's face reality. Farms today are probably too big and they need to shrink some, but we're not going to go back to the small-scale operation. I just don't see it happening."

Gabe explains that while his ranch is average in size for North Dakota, in general, people going into regenerative agriculture can do well economically at a smaller scale. "If you're practicing my type of regenerative ag, you could easily do it on a thousand acres. Farmers could be much, much smaller and we'd get more people out on the land. We could sell more local foods and that's a good thing."

Like many people in the regenerative agriculture movement, Gabe's goal is to be both organic and regenerative (aka regenerative organic). He and others around the world are working to achieve this new gold standard of agriculture but it's tricky. Gabe says that in order to achieve regenerative organic on a large operation, they must perfect the correct sequencing of crop rotation and animal impact in order to completely eliminate the use of herbicides. He believes he is getting closer to doing this. "I'm not going to give up," he says.

PASS THE BATON

In every sense of the word, Brown's Ranch is a family operation. Gabe and Paul work together on a daily basis. Their houses are about five miles apart. Gabe makes no bones about it; his mission is to pass the ranch on to Paul. While not an overly emotional man, Gabe brims with pride when he mentions this. After he came back to the operation from college Paul, who is now almost thirty, lobbied to expand the work and the products they offer.

"Working with my father is interesting," says Paul, laughing. "It's enjoyable. It really is." Gabe, who is listening nearby, says that in spite of his eavesdropping, Paul should tell it like it is. "I went to college for four years, and I can count on one hand the number of my friends that actually went back to the farming or ranching operations. They blamed it on 'there's not enough money for us to come back. There's not enough work for us.' Well, if farms and ranches would diversify, there're plenty of opportunities."

I ask Paul what his college education lacked so that his friends did not enter farming. "I think what's missing from the education system is practicality. I can count on two hands the amount of times we went out on the land for classes. And that's a big disconnect." He tells me that most of his friends who did go back to farming and ranching followed the conventional industrial model taught at the local agricultural college. They spend their days managing inputs, which must be purchased, and then managing the grain, which must be trucked to the grain elevators (where it is counted to be sold on the futures market on Wall Street). "They just spend so much time trying to get ahead. And when you're in that kind of conventional model, you waste your whole life just chasing your tail. So it is sad to see some of my friends going down that path," he laments.

In contrast, by working together, the Brown men have increased the ranch's effectiveness and income streams. "Since coming back from college and joining my dad on the operation, we've easily increased the net profit of this whole ranch by fifty percent," says Paul. "The ranch has been debt-free for quite a few years now. The profit comes from the wealth of the land, and if the land is in a healthy state, you don't have to spend so much money on all these inputs that act as crutches to take the place of a healthy landscape." Like father, like son.

Unlike the average person from his generation, who graduates with $30,000 in debt (mostly to the US government in the form of loans), Paul's start in his career is quite different. "My debt, personally, is none. So the ranch helped pay for my education. Yeah, it feels good to start out on a level playing field and only go one way, hopefully up."

Millennials take note: It may be time to consider a career change. Today there are only around 120,000 farmers aged thirty-four and younger in America. Meanwhile, the number of farmers sixty-five and older is steadily increasing, from around 558,000 in 2002 to 701,000 in 2012, and the trend continues.[3] During that same time, the number of "beginning farmers" (those with less than ten years' experience) plummeted by 20 percent. Each year, more farmers reach retirement age and fewer new farmers show up. As older farmers "age up," they are leaving in their wake a void that could be filled with young, enthusiastic, well-trained regenerative farmers like Paul Brown.

In addition to managing livestock and putting away hay for the winter and doing general work around the ranch, Paul is in charge of their online business. He answers customer e-mails, works on their ordering system, and is constantly looking for ways to use the Internet to increase their outreach. His outlook on the future is bright. "One day, I'd like to have children to take over this operation, and I'd like to develop a relationship with my sons and daughters like I have with my dad and pass this ranch on to the next generation," he says.

When I ask him if he has a girlfriend, he tells me he has a few "prospects." His father chides him, saying Paul works too hard and needs to instead spend a bit of time dating. Paul answers defiantly, "It'll happen. I'm not even thirty yet."

APPLES TO APPLES

Gabe is taking me on the final piece of his farm tour. We head several miles on dirt roads turning left, then right, then straight, then left. Like so many farmers, Gabe must have a map somewhere in his cavernous brain of where all his fields are versus where his neighbors' fields are. With five thousand acres split into bits and pieces, that must be a heckuva map. Finally, he pulls off to the side and parks the truck in a field that looks like it was burned by brushfires.

He gets out, walks around to the back of his truck, and leans on it, surveying the dead field. "Instead of growing a crop on here this year, this producer decided he would just do what's called chemical fallow. In other words, he sprayed herbicides on here to kill the weeds, and now the land is idle. He's doing that thinking it'll store moisture, but what he doesn't realize is he's degrading the soil even further by not having a living plant on here," says Gabe.

He continues pointing into the field. "The irony is, if you look out across his field, you'll see those are Roundup-resistant weeds. So he's just perpetuating his weed problem by doing this." The only thing that looks alive in the field are clumps of marestail, a tall, green, furry-looking weed (and one that is becoming increasingly resistant to Roundup across the country).

Gabe says, "Nowadays, glyphosate is used like water." He's talking about the primary chemical component of Roundup, which by itself makes up a big portion of the three pounds of pesticides sprayed per American every year. "They apply it in multiple passes, and glyphosate being a chelator, it ties up the heavy metals so there's less nutrients cycling. Then you have deficiencies in your plant so the plant isn't able to ward off fungal diseases."

Gabe bends down, picks at the dry dirt, and manages to grab a clump in his hand. He brings it to his nose, sniffs it, and shakes his head. "And if you smell the soil, it's highly bacterial, which means it's bacterial-dominant." He offers it to me and I sniff it. Gone is that earthy smell from Gabe's garden. In its place is a sickly, acidic odor. "Look at the density of that," he says, crunching the hard, compacted soil in his hands. "You see no earthworms in there, no holes. It's gonna have very poor infiltration, very little nutrient cycling going on. It's more or less just dirt, not soil," he says, tossing the "dirt" back on the ground. The lines on Gabe's brow are deep and furrowed, like a plowed field. He's visibly distressed.

Gabe has brought me to this particular "field of death" because it offers a side-by-side "apples to apples"–type comparison. Right across the dirt road is one of his polycrop polycultures, a veritable plant forest brimming with life, including the loud crickets. I ask Gabe to dig a shovelful of soil from each field. He obliges.

The dirt from the death field looks like hardened clay with no roots. The soil from Gabe's field is like chocolate cake. He says it takes between three and five years for him to restore an "extremely degraded" piece of land like the one we are standing in back to basic soil health.

The more I look at the two fields, the more I realize we are standing on the fence line between two futures. In one of those futures a rapidly expanding global population will pressure farmers to add more chemicals, more engineering, more machines, and more fossil fuels to produce more nutrient-void, chemically laden calories. The other future is part of a much more complex ecosystemic way of plugging into the flow of natural processes. This alternate future involves relearning our food system and then redesigning it from scratch. It also offers the promise of foods that just might have the potential to save our civilization from self-destruction.

We stand there in silence for a moment, looking at both fields. Gabe sums it up by slowly pointing to each field. "I ask you: Is this gonna feed the ten billion people?" he says, jabbing his thick finger toward the field of Roundup-resistant weeds. "Or is this?" he asks, opening his palm to his polycrop forest of food.

To anyone bothering to come to North Dakota to look, the answer is painfully obvious.

THE GREAT NORTHERN SUN

It's time to leave. I'm chatting with Gabe about his bottom line. He shows me a diagram on his laptop of all their income streams. There are numerous "ins" and not a whole lot of "outs" in terms of money.

He says, "Sometimes when I give presentations I get this group that just says no way, that can't happen here. So I offer them this challenge. I say, 'I will bet my ranch against yours that I can get it to work on your operation.' Because the principles of soil health—the least amount of mechanical disturbance, diversity, armor on the soil, living roots at all times, animal integration—those principles are universal everywhere in the world where there's production agriculture. So there is zero doubt, and I mean zero doubt, in my mind that I can get these practices to work anywhere in the world. And I'll take that bet with anyone who wants to place it with me."

He has one last comment before I hit the road. "At the end of those four years of drought and hail, we were dead broke. I mean, I joked that the banker knew when we bought toilet paper, we were so broke. And now we can hand the ranch over to our children debt-free, and we have enough money to retire, and I'm fifty-four years old. So it's highly profitable in this type of system. If you farm and ranch in nature's image, nature will take care of you. It's as simple as that."

As Brown's Ranch fades in the rearview mirror, I think back to Gabe's comments on the government insurance–industrial farming catch-22 in which most producers find themselves. The bastardization of a program that was originally intended to help farmers now ensures that the market is viciously opposed to farmers' success. By institu-

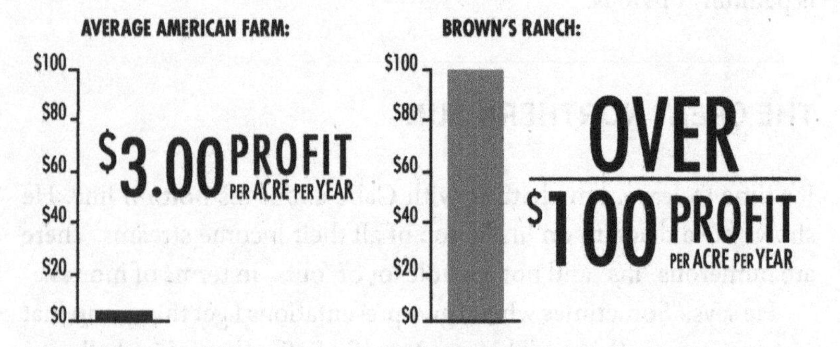

If America's nine hundred million acres of production made as much per acre as Brown's Ranch, US agriculture would make $90 billion a year instead of costing taxpayers money.

tionalizing the overproduction of carbohydrate-rich commodities, most of which are fed to feedlot animals, our own government has locked medium- and large-scale farmers into a game they cannot win.

Gabe believes the key to breaking free from the viselike grip of the commodity game is a new relationship, a new bond, between food producers and consumers. The $43 billion-a-year organic foods industry is certainly evidence of the growing demand for such a relationship. But to truly revolutionize agriculture, it will take no less than a herculean effort from both the farmer/rancher and the consumer. Like all movements, this will not and cannot start with entrenched government policy. In fact, regenerative agriculture must, at least for now, fly in the face of it.

Ultimately a vanguard of concerned consumers and maverick farmers and ranchers must lead the charge toward regenerative agriculture. Knowing that there are farmers and ranchers like Gabe who can meet demand is important. But making this shift ultimately rests with consumers. With the food habits of 98 percent of the population dictating what the other 2 percent grows, it's up to us, the eaters, to change the system that brings us nourishment.

It's time for us to redefine *food*.

CHAPTER 9

A NEW PLATE

With a few exceptions, chefs were once thought of as sweaty, replaceable laborers. But not anymore. The rise of TV-ready and brand-friendly chefs has coincided with something Americans are doing a lot more—eating out. Since 1970, restaurant food sales have skyrocketed from around $43 billion to over $520 billion.[1]

Of course, all this eating out has to do with something we are doing a lot less—cooking at home. Surprisingly, this trend cuts across economic sectors, with low-, medium-, and high-income households all eating less at home each year. Even when we do eat at home the food on the dinner table is increasingly not prepared there. Today, less than 60 percent of dinners eaten at home were actually made at home.[2]

All of this has fueled the fires of a new breed of chef who has their own TV shows, food products, Vegas restaurants, blogs, and epicurean followings. It has also made some of them extremely wealthy. A look at the top ten celebrity chefs puts their combined net worth at almost $2 billion. Not bad for a bunch of cooks.

With hundreds of television shows about cooking, movies like Jon Favreau's 2014 *Chef*, whole networks now dedicated to nothing but food, and an Internet culture obsessed with photographing everything it eats, it is odd there is so little time and attention spent on where that food actually comes from, who grows it, or how it is grown. This is why I'm driving north from New York City to meet a celebrity chef

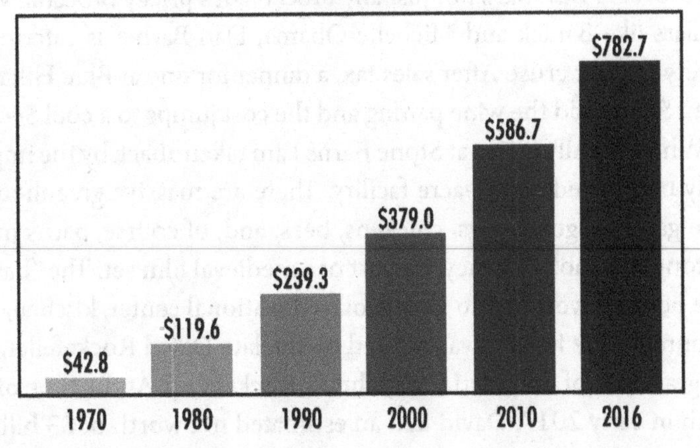

Americans are eating out a lot more. Even when they eat at home, they often eat takeout. All of this contributes to the boom in restaurant sales. *Data:* National Restaurant Association.

whose *New York Times*–bestselling book *The Third Plate: Field Notes on the Future of Food* and popular Blue Hill restaurants aim to turn food upside down.

For a chef not even listed in the top fifty "celebrity" chefs, Dan Barber is a hard man to get to. Maybe that's because of all the hyped-up, flame-throwing, gourmet-producing creators he's the only one who is attempting to make food that is part of an ecosystem.

At his Blue Hill restaurant at the Stone Barns Center for Food & Agriculture in Tarrytown, New York, food is grown literally outside the front door. Dan has become famous for doing some pretty audacious things, including parading a whole dead hog, hogtied to a spit, through his restaurant to educate his customers about where their meal is coming from. He has served food on dirt—literally. And then there was the time he served people expecting a full meal half a head of braised cauliflower. Yep, these are the type of things that make a chef a celebrity.

Oh, and his appearance on the popular Netflix show *Chef's Table* didn't hurt, either. The show encapsulates Dan's theory on food and makes the viewer salivate for the simple things in life. Things like carrots and broccoli. But that's not just any broccoli, it's pricey broccoli. With regulars like Barack and Michelle Obama, Dan Barber is catering to society's upper crust. After sales tax, a dinner for one at Blue Hill runs about $245. Add the wine pairing and the cost jumps to a cool $449.

When I finally arrive at Stone Barns I am taken aback by the impeccably manicured eighty-acre facility. There are massive greenhouses, huge gardens, goats, pigs, chickens, bees, and, of course, barns made of stone that look like they are part of a medieval film set. The "barns" have been converted into a gorgeous educational center, kitchen, and restaurant. The facility was created by the late David Rockefeller, the last grandson of Standard Oil's John D. Rockefeller. At the time of his death in early 2017, David had an estimated net worth of $3 billion. It's nice to know that some of that oil money is dripping down to reform food and agriculture.

I am met by an assistant who used to work in marketing in this company and that in New York City. She briskly shows me around, keeping a tight eye on the schedule. When the tour of the gardens, vast greenhouses, and regenerative farm concludes, I am delivered to an antechamber next to the restaurant. This odd room, stacked with chairs and tables, must be used occasionally for private dinners. Another assistant, this one even more tightly wound than the first, takes over. She advises me in a breathless explanation that the day is running "terribly behind schedule" and that Dan may not have time to meet, but I should wait here just in case.

I sit and wait. An hour passes. Still more time passes. Restaurant staff goes in and out of the room, carrying various items and folding napkins. Somebody must have taken pity on me, because out of nowhere a tray of soup and sandwiches appears. To say the food is delicious would be an understatement. The food is instead ridiculously scrumptious, and that is still not doing it justice. The meal definitely made the wait a little more digestible, but I'm still hoping to meet Dan by the time I finish my last bite.

Finally, Dan arrives with his assistant, who informs me he's got fifteen minutes exactly—not a second more. We sit, and I begin to ask questions. As Dan speaks in his rapid-fire New York accent, he reminds me of a Seinfeld impersonation of an Upper East Side chef. But he's also sincere and super-passionate about food.

I begin by asking him how he got started. "I had a farmer that I was particularly drawn to because he was growing the most incredible wheat I had ever tasted. We were baking bread with it and it was stunningly delicious. I went to this farmer's farm and stood in the middle of his field and I'm looking at buckwheat and barley and beans and lots of cover crops but I see almost no wheat," says Dan. "What was explained to me is that all the crops are planted with meticulous thought about their rotation potential in and around the farm to create the kind of soil conditions that enabled me to have the wheat I so coveted. But the soil-supporting crops were not crops I was supporting in my kitchen."

Dan tells me that this was his big aha moment. "I thought here I am, a 'farm to table' chef, hanging my hat on this direct connection with a farmer but I am skimming the cream. I'm eating high on the hog," he says. "I think that's the problem with 'farm to table.' We talk about 'nose to tail,' of eating the whole animal to respect the animal, but do we ever talk about nose to tail of the whole farm?"

Among Barber's signature dishes is his "Rotation Risotto," which is made with all the grains that support the wheat crop, including millet and buckwheat. Dan presented the risotto, which is still on his menu today, as part of a pop-up restaurant in New York City called WastED. The temporary restaurant was designed to create food from, and educate customers about, things that are usually discarded in the process of making food.

Barber says that we have to move toward "a way of eating where the region dictates what you should be eating not the other way around." He says that if you survey the history of food globally, food itself really comes from what the land has the ability to grow.

"Look at the history of cuisines and patterns of eating for different cultures," he says. "We think of Japan as a rice culture but you know in order for the rice to grow, buckwheat had to be the rotation crop into rice. So they created soba noodles. You ate your rice but you also ate your soba noodles. You had to do that but it wasn't forced on you; it was just part of the culture and it was delicious. And in India it's lentils, a leguminous crop for all the other crops. In North Africa it's millet."

Dan Barber says that while traditional cuisines emphasize eating soil-supporting crops, they also de-emphasize eating meat. That's where Barber has used his celebrity credentials to draw as much attention as possible to what he calls "the third plate," a plate that is neither meat-centric nor vegetarian but rather vegetable-centric with a bit of meat garnish.

"To a lot of people, it's quite shocking to think about their dinner without a center cut of protein—whether it's a chicken breast or a steak or a leg of lamb or a honking piece of salmon," he chides. "This protein-centric ideal that the Western diet is famous for has to be reversed. If you think about a Cantonese meal, you see some rice and there's meat, but it's a smattering of meat. Indian cuisine's the same way. Italian cuisine actually is very much the same way. Peasant French food is the same as well."

Dan Barber's idea of a new plate is a large center cut of vegetables or a vegetable-based dish with meat sauce or meat extras. His "rebel" cooking style has inspired a dedicated following of foodies who blog religiously about his culinary achievements. His dishes regularly feature things like a huge carrot atop some meat sauce or a smattering of various veggies with two hard-boiled eggs. He makes "normal" dishes too, including lamb, scallops, and the like. But even these dishes generally incorporate some of the lesser-known parts of the beast.

While Barber is making waves in the foodie community, his idea of the third plate is directly in line with some of the best medical science on earth.

THE HEALTHY PLATE

In 2011 the Department of Agriculture replaced its age-old food pyramid with a simple round diagram called MyPlate. In some ways the plate graphic is a step toward promoting health. It calls for 50 percent of one's diet to be derived from fruits and vegetables. But in other ways, MyPlate is deficient. It calls for an unnecessary helping of dairy with each meal. Other than "varying" protein sources and "whole" grains, MyPlate offers no guidance in terms of where the 30 percent grains and 20 percent protein should be sourced.

Later that year, the Harvard School of Public Health (HSPH), acting with Harvard Health Publications, unveiled the "Healthy Eating Plate." The impetus was to correct MyPlate. According to Walter Willett, professor of epidemiology and nutrition and former chair of the Department of Nutrition at HSPH, "Unfortunately, like the earlier US Department of Agriculture pyramids, MyPlate mixes science with the influence of powerful agricultural interests, which is not the recipe for healthy eating."

What Willett didn't quite say is that despite any perfunctory attempt by the US government at promoting healthy eating, Washington and the Corn-CAFO complex are joined at the hip. The numbers don't lie. Thanks to the federal crop insurance program, 80 percent of America's crop acreage is planted with just four crops: field corn, soy, hay, and wheat. Consequently, 63 percent of the average American daily caloric intake is derived mostly from those crops or from animals that ate those crops.

Consider that America has about 915 million acres of range and farmland, of which about 330 million are planted with "principal crops," including mostly field corn, soy, hay, and wheat. But only about 7 million acres, or 2 percent, is planted with "pulses and vegetables," including things like carrots, lettuce, and broccoli. The result is that the average American only gets about 12 percent of his or her calories from vegetables, and most of that from potatoes (which the Healthy Eating Plate guidelines specifically say "don't count").

What we plant dictates what we eat. And at least in America, the sneaky way of ensuring a commodity agricultural system that requires toxic chemicals is to promote a diet based on those crops and chemicals. Supply must, after all, equal demand. What is shocking about the Standard American Diet is how each layer in our food system is interlinked with the next.

Here's how it works: The middle of our country grows bulk commodities of mostly empty calories while fresh stuff is grown on the coasts. Like our country, our grocery stores and fast-food joints have the bulk food in the middle and fresh things around the edges. Our diet is, again, a reflection of that, with the majority of our calories coming from the nutritionally void bulk in the "middle." Not surprisingly, the average American is twenty-three pounds over his or her ideal body weight, most of it from the bulk in the middle. And around our middle is where that jiggling bulk, in all its glory, ends up.

Then there's the secret ingredient that makes the whole machine work.

Pick up a box, bag, or carton of something in the supermarket and scan through the ingredients. Chances are it will be there. Often hidden by names like "crystalline fructose," "glucose syrup," or "dahlia syrup," its real name is corn syrup. This sweet substance has found its way into the majority of what we eat and, along with other processed sugars, now accounts for around 14 percent of the average American's daily caloric intake.

The hidden function of corn syrup and other refined sugars is to sell the megacrops of corn, soy, hay, and wheat at a high markup. These crops are processed through a food system that delivers them to us in brightly colored packages filled with addictive flavors. Thanks to sugar, which tricks our taste buds and our brains, Americans have been sweetly seduced into eating a diet that is in every way antithetical to our health and the health of the ecosystem.

Our entire chemical-industrial agriculture system, our Corn-CAFO complex, our GMOs, our commodity crops, our federal insurance program, and even our government's nutritional guidance have delivered unto Americans a state of physical being that is aptly described by the title of the 2010 documentary *Fat, Sick and Nearly Dead*. As the incidence of diabetes, heart attacks, and other nutrition-related disorders skyrockets, we must begin to see the destruction of our bodies and our ecosystem as one and the same. Our modern food system has not delivered on its promise of food security. It is instead trying to kill us.

In contrast, Harvard's model gives very different diet advice which, instead of enriching chemical manufacturers and grain traders, is based on some of the best medical science on the planet and backed by decades of research studies.

Here are the guidelines set forth in the Harvard Healthy Eating Plate:

STANDARD AMERICAN DIET

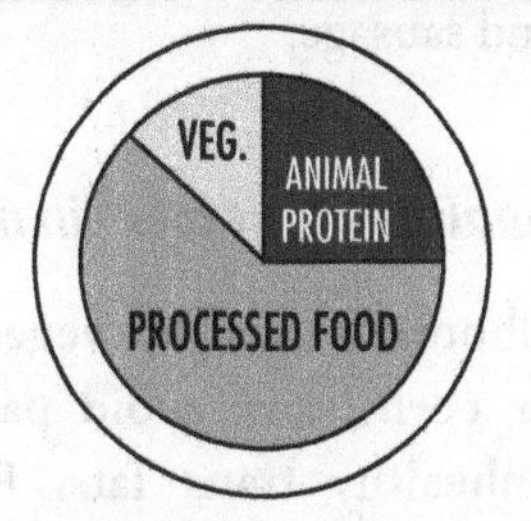

USDA's "MyPlate"

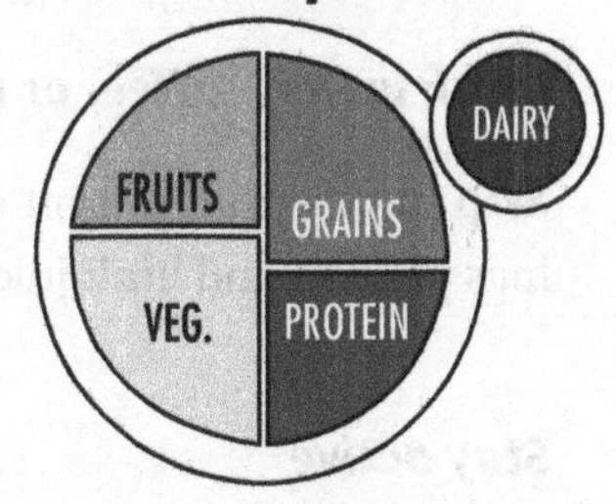

HARVARD's "HEALTHY EATING PLATE"

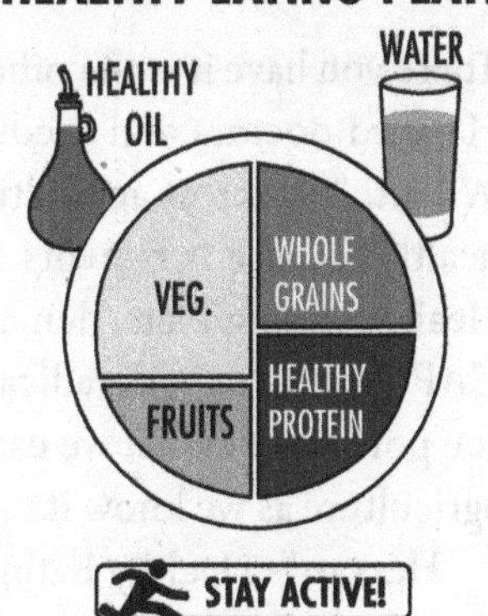

Make most of your meal vegetables and fruits—½ of your plate

Aim for color and variety, and remember that potatoes don't count as vegetables on the Healthy Eating Plate because of their negative impact on blood sugar.

Go for whole grains—¼ of your plate

Whole and intact grains—whole wheat, barley, wheat berries, quinoa, oats, brown rice, and foods made with them, such as whole wheat pasta—have a milder effect on blood sugar and insulin than white bread, white rice, and other refined grains.

Protein power—¼ of your plate

Fish, chicken, beans, and nuts are all healthy, versatile protein sources—they can be mixed into salads, and pair well with vegetables on a plate. Limit red meat, and avoid processed meats such as bacon and sausage.

Healthy plant oils—in moderation

Choose a healthy vegetable oil like olive oil (not canola, soy, or corn), and avoid partially hydrogenated oils, which contain unhealthy trans fats. Remember that low-fat does not mean "healthy."

Drink water, coffee, or tea

Skip sugary drinks, limit milk and dairy products to one to two servings per day, and limit juice to a small glass per day.

Stay active

The red figure running across the Healthy Eating Plate's place mat is a reminder that staying active is also important in weight control.

There you have it—the official advice on the optimum human diet by Harvard doctors and medical professors. As indicated by Professor Willett, "powerful agricultural interests" have thwarted our view of health because it benefits their bottom line. If America follows the Healthy Eating Plate, demand for the commodity crops of the Corn-CAFO complex will radically diminish. In other words, just shifting the portions of what we eat away from processed junk would change agriculture as we know it.

Harvard's Healthy Eating Plate spells out the basic portions of the very thing that is the key to creating the chain reaction that builds soil and sequesters CO_2: a regenerative diet. Currently missing from the Healthy Eating Plate is one line of text: "Do your best, your very best, to make sure the items on the plate are sourced from certified organic, or better yet, regenerative organic farmers." It's possible that with time and growing awareness, those words, or something similar, will one day be added. (After all, what self-respecting health scientist would willingly advocate for eating food that contains chemical poisons?)

The simple truth is that if we want to shift the world of agriculture and how it impacts our soils, we first have to shift what's on our plates. And if we want to pull the CO_2 that's in our atmosphere back into the ground, we darn well better make sure that our food is sourced regeneratively.

EATING DIRT

Back at Stone Barns in bucolic upstate New York, my fifteen minutes are almost up.

Dan Barber says that critical negotiation between the culture and the ecology of a landscape was never present in America because ours is a relatively young agriculture that is based on some of the richest soils in the world. Considering American agriculture began in earnest less than two hundred years ago with deep, dark soils created by millions (and over time, billions) of herd animals that once roamed our land, we have done a marvelous job in a short period of time of stripping its fertility. Says Barber: "When soil fertility started declining on the East Coast we moved west to rip up virgin soils and get that productivity that we had become so known for."

Dan is all about leaving a place better than we found it. "Agriculture . . ." He says the word like it's a sentence in and of itself. "For sure it's a disruption of the natural system. There's no getting around that. But can we, through our diets, disrupt a natural system thoughtfully and elegantly and lightly and actually if we really do it right, if we really do it right, can we improve our natural system? Phew. Can we? I mean, that's kind of a haughty thing to say. A natural system just seems so perfect—it's nature. But actually there's ways to think about agriculture that improve an ecological place and make it a healthier and more delicious possibility for food production."

Barber believes that the "midsized" farms of between two and five thousand acres (farms like Gabe Brown's) represent the biggest opportunity for change. He says that these farms, which are currently failing very quickly, are too big to mess around with farmers' markets, but too small to compete in the commodity game.

"What's ironic," he muses, "is that these are the very farms that are most nimble. They're most able to adjust to a diverse diet. In fact, they could be very easily motivated to grow a whole host of diverse crops that are soil supporting and delicious. But the diversity doesn't come from the farmer. It cannot. It has to come from the culture. It has to come from the demand and that means demanding not just a direct connection with a farmer, but a diverse diet that responds to the ecological necessities of the landscape.

"When we eat we are eating soil fertility," the chef continues. "A delicious plate of food is a window into a healthy landscape. By definition you can't have delicious food without something going very, very right in environmental stewardship and soil health. From a flavor perspective the best way is through a rotation of intelligent crops. And our job as chefs and our job as eaters is to figure out how to make that coveted and delicious. How do we do that in America where all we know is this abundance?"

This, it seems, is a rhetorical question. Dan Barber has already spent his entire career working on the answer. "The blessing of the American food culture is the lack of culture," he contends. "Because of our buying power and because of our numbers. We move with rapid speed and we can change landscapes quite rapidly. That's one thing Americans do very well is going after and paying a little bit more for something when they feel greedy for it. So we've got to create that greed and be greedy for the right things."

No sooner has Dan finished his last sentence than the assistant comes in to tell me his time is up. I ask Dan to sign a release for me to use his image and quotes, which he does happily. He's whisked out of the room to attend to an army of white-clothed chef's assistants. A few moments later the assistant comes in with the release in hand, which she tears up in front of me, explaining that they never give permission for anything of the sort without first reviewing the finished material.

Stunned, I pack up my belongings and begin to head out. The assistant must feel a pang of remorse. She appears just before I depart with a loaf of freshly baked bread, made, of course, from that very special wheat that Dan supports with his rotation risotto. Inside its custom-made cotton cozy (yes, it has its own cozy), the bread is warm.

I call my office to check in and say that I have scored a loaf of Dan Barber's bread. By the time I am sitting on the tarmac at LaGuardia Airport, word has gotten around that I have a $100, no a $300, um, rather a $500 loaf of bread. I am getting texts asking if I can bring it home to share with various friends and family. It's too late, though. The bread, delicious and warm as it was, is long gone. Like any scrumptious grain product, its addictive effect *made* me eat it all before I even realized what was happening.

For all the celebrity chef–driven zany behavior around Dan Barber, he knows how to make simple foods taste great. His vision for high-end, epicurean-driven, agricultural reform may not be that crazy when you consider how Elon Musk created the Tesla first for luxury car buyers. Only after he proved there was a demand for his product and, after he made the company worth billions, other carmakers perked up and began to put plugs on cars for the rest of us.

The big difference between a Musk-like revolution and an agrarian one is that the barrier to entry is much lower with agriculture. To farm you need land, some scraps of startup capital, and a heckuva lot of endurance. With luck and a bit of research, you can grow food. And, as farmers and ranchers like Gabe and Paul Brown have shown, if you're committed and clever, you can grow a lot of it and make money, too.

However, the true test of Dan's model of demand-driven agriculture change isn't in fancy restaurants. It's in places that cater to the bulk of eaters. Places like a certain café chain in Southern California.

3 SQUARES

Ryland Engelhart is an affable thirtysomething-year-old. With his slender frame, hair cropped to the scalp, constant five-o'clock shadow, always-on-iPhone, and smiling face, you'd be hard pressed to pick him out of the crowd at a Wilco concert.

I first met Ryland several years ago when he and his wife, Sarah, were in the process of moving to Venice. They were looking for a place to live. Meanwhile, my wife and I were moving out of the city. They were starting their new life and we were leaving our old one. Ryland and Sarah ended up moving into the house we had rented for a decade and they more than filled any void we left behind.

For the last several years, Ryland has been banging the drum of soil by holding events, sending out newsletters, and getting a nonprofit together to raise awareness about the importance of soil. Today that nonprofit is called Kiss the Ground (the inspiration behind this book's title). In many ways, he was the inspiration for me to take this journey. And now I've come full circle.

I meet up with Ryland early one foggy Venice morning. Sitting with Ry, as his friends call him, in the garage I once used as my office makes me feel like I'm in a time warp. In my wild and crazy twenties when the sky was the limit and anything was possible, I too had moved to Venice with dreams of making the world a better place. Now it's Ryland's turn.

"My title is CIO of Café Gratitude, which is Chief Inspiration Officer," explains Ryland. "Café Gratitude is a family business. Started by my father and stepmother about twelve years ago. And since then, my brother and I have joined and now run the new Southern California tier of the organization. We have three restaurants here in Los Angeles. We have two Café Gratitudes, one Gracias Madre, and two more Café Gratitudes in Orange County."

Notice he didn't say "joined the restaurant chain"? That's because for Ryland and his family, the cafés really are an extension of a set of organizing principles. Ryland says, "When we started Café Gratitude, it really was to demonstrate a new business paradigm. Ultimately, it was about 'how do we improve life on Planet Earth through a very localized model?' "

The Engelhart family enterprise got off to a rough start in the San Francisco Bay Area. The restaurant, which had begun in 2003 with one small café, quickly grew to include eight locations. But after being criticized for "forcing" employees to attend a self-help seminar called the Landmark Forum and an ensuing public legal dispute related to employee tip sharing, in 2011 they closed all their Bay Area restaurants. On the Café Gratitude Facebook page, Ryland's stepmother, Terces Engelhart, cited the steep cost of organic produce, small margins, and heavy legal bills as the causes of the closures.

Having learned from their failures, the Engelharts' new crop of mission-driven cafés now seem on track for rapid growth. "We started with fourteen employees on Twentieth and Harrison in San Francisco. And now, just in Southern California, we have over five hundred employees," says Ryland.

Arriving at Café Gratitude's Venice location, I open the door and am greeted by a young woman with a bright smile and a clean white V-neck T-shirt with gold letters that, when you count the *V* of the V-neck, spell *LOVE*. She shows me to a table on the porch outside. From here I watch the sidewalk menagerie of Venice hipsters.

A waiter brings me a menu and tells me a little about the restaurant. Everything is 100 percent vegan and 100 percent organic. Across the top of the menu in huge letters that look like the first line of an eye exam are the words "I AM . . ." followed by the smaller names of menu options, each of which is an affirmation. There is the "ADVENTUROUS" coconut curry raw soup and the "FABULOUS" raw lasagna. There's even a "GRATEFUL" bowl of kale, rice, and beans for which you can pay as little as three dollars (if you're pressed for dough) or as much as you'd like (in order to donate to the meal of somebody you'll likely never meet).

I order the "I am COMFORTED" appetizer of roasted yams with rosemary. I add an "I am EXTRAORDINARY" house "BLT" wrap made with coconut bacon. To round it out, I get an "I am COZY" hot drink made from lemon, ginger, honey, and cayenne. I resist the desserts (for now) but the "I am DELIGHTED" raw chocolates and truffles have been seducing me ever since I walked past the dessert cooler on the way in. There's also the "I am MYSTICAL" raw cacao and coconut macaroon.

As I order each item, my waiter repeats my affirmation, "Okay, you are COMFORTED, you are EXTRAORDINARY, and . . . you are COZY." I skim down to the bottom of the menu. The paragraph I find in place of the usual "we include organic ingredients when possible" says:

> Café Gratitude is our expression of a world of plenty. Our food and people are a celebration of our aliveness. We select the finest organic ingredients to honor the Earth and ourselves, as we are one and the same. We support local farmers, sustainable agriculture, and environmentally friendly products. Our food is prepared with love. We invite you to step inside and enjoy being someone who chooses: loving your life, adoring yourself, accepting the world, being generous and grateful every day, and experiencing being provided for. Have fun and enjoy being nourished.

It's not political (exactly) and it's not spiritual (exactly), but it certainly has undertones of both. One could definitely say it's a bit hippy-dippy but then again this is the only 100 percent organic, 100 percent vegan restaurant chain in America. And like the organic movement itself, this is a restaurant chain that is growing by leaps and bounds.

Serving a thousand people per day per restaurant at three locations in Los Angeles, the Café Gratitude philosophy definitely appeals to cost-conscious foodies. Before the waiter departs to put in my order he asks if I'd like to hear the question of the day. I say sure. He asks me, "What brings you peace?" He says that since I'm dining alone, I can share my answer with him, just ponder it, or share it with people in my life.

I try to let the whole experience penetrate my cynical Gen X mind.

In some ways, it's no surprise that Ryland and his family have been criticized both online and off as being part of a "cult." They are certainly at the vanguard of a movement that combines a certain brand of spirituality with a type of food ethics.

At the Venice café there is a small shelf to honor an employee who suddenly passed away. There is a picture of Jesus, a Mother Mary candle, a small painting of the Hindu god Shiva, and a photograph of a swami. There's also a crystal and some incense. It's a genuine Venice altar and a testament to the Engelharts' "come one, come all" attitude.

Having arrived at the restaurant and managed to speak to and smile at just about everyone he meets along the way, Ryland sits down at my table to see how I'm enjoying my lunch. I ask him how the restaurant started.

Ryland says the idea of the restaurant actually originated from a board game that his parents created called the Abounding River. "It was just a game where people would travel around a board like in Monopoly or Life," he explains. "And they would pull cards and those cards would give them an opportunity to look at their life and see where they're seeing the glass half empty or half full. So the whole idea of the restaurant was to create an environment that elicits this expression of gratitude, fulfillment, love, kindness, and generosity. The question was: Can we create a restaurant that will create hundreds of ripple effects?"

Ryland gets up to attend to something. Like everyone eating alone in LA, I use my smartphone to make myself appear busy. I decide to order an Abounding River game. But this turns out to be more difficult than I thought. I can't find the game new or used anywhere for under $80. So I try eBay. There, shrink-wrapped, "NIB" (New in Box) is an Abounding River Board Game up for auction starting at $20. I place a bid of $21 with a secret maximum of $25, forget about it, and get back to my baconless BLT.

The next time I grab a few minutes with Ryland he explains that the game formed the basis for the type of food they would serve. Since their game was about making the world a better place, they felt their restaurant had to serve healthy fare. "For eleven years, we've been one hundred percent organic ingredients all the time. Most people say, 'Organic when price permits.' But price never permits. It's always more expensive!" Ryland adds, laughing.

Walking their talk has made the Engelhart family rethink their own assumptions around food. "Because we live in cities and food is grown in the country, we have no relationship to what we eat. We are completely disconnected from what the food system looks like," says Ryland. "There's no longer an agrarian culture, a connection to our food." The Engelharts are trying to use their restaurants to rebuild that connection.

The items on the Café Gratitude menu change with the seasons. The spaghetti squash noodles aren't available in summer and you can't get the coconut soup in winter. Ryland says that every decision for the menu depends on how those foods will affect the farmers upstream. That's because Ryland and his family believe that the demand side (restaurants and grocery stores) drives the supply side (farmers and agriculture).

Like Dan Barber, Ryland says we have to create a craving for what is easy, economical, and sensible for the farmer to grow. We must start with the soil, then look to the farm and finally to the table, not the other way around. "It can't just be 'we wanna eat what we wanna eat when we wanna eat it,' " he says.

All of this is what led Ryland's father to start their own farm to feed their restaurants. That farm is an experiment in what Ryland says is "the cultivation of the kind of food that we need for the future." In other words, regenerative agriculture.

A quick recap: Regenerative agriculture builds soil fertility. It's not "sustainable" in that it doesn't sustain the current status quo. Instead, it is designed around the principle that, with proper management, nature can be additive. Only nature (rather than man-made chemical inputs) can increase the true "holy trinity" of carbon, nitrogen, and water in the soil while simultaneously producing copious quantities of food. To do this either requires compost or animal husbandry or both.

Compost is part of most small-scale organic farms. But like the farmers of Brittany and Rudolf Steiner, today's regenerative farmers are finding manure is the secret sauce when it comes to building soil on a large scale. Just as nature does not occur without both the plant and animal kingdoms, so too are regenerative farmers finding they must embrace both life-forms. This embrace is creating some challenges, especially for vegetarians like the Engelharts.

BE LOVE

Vacaville, California, draws me back like a dust-riddled magnet. This time I'm not headed to a garbage dump turned compost facility but to a small-scale organic farm run by Ryland's father and stepmother.

Located about five miles out of town, Be Love Farm is sequestered away from the chemicals and industrial agriculture of Central California in a little place called Pleasant Valley. As I drive up I am struck by the plethora of green that surrounds the farm. It's no rain forest, but compared to the dirt wasteland that separates Vacaville from Sacramento and San Francisco, this place pulsates with life.

The sun is starting its long journey toward the horizon but the day's heat still hangs oppressively in the air. Ryland's father, Matthew, suggests I take a swim to cool off. From the small pool I can see an adobe-style main house that is still very much under construction, a couple of brightly painted outhouses, and an old camping trailer that has become part of the landscape. Chickens roam freely, pecking at the ground. A man walks by leading a couple of cows with bridles.

After my swim it's time for dinner. When I walk into their home, Terces Engelhart, the co-owner/operator of the farm and Matthew's wife, is working in a large semi-industrial-sized kitchen and has begun to put food on a counter. Matthew gathers all those present on the farm and we stand around the table and hold hands. I follow their lead as we go around in a circle, taking turns saying something we are grateful for.

Dinner is simple fare: salad, bread, cheese, homemade salsa, and some pasta. Aside from the pasta and the bread, the ingredients have come from the farm. Over dinner Matthew, who is fifty-eight, explains his vegetarian philosophy. "I haven't eaten flesh since I was seventeen, except for very occasional fish." He says his choice to become a vegetarian was a combination of things. "There was the spiritual aspect, health aspect, and environmental aspect."

Matthew says he was a raw foods vegan for some time but felt he wasn't getting the needed body warmth. He switched to eating a cooked but still vegan diet. Having lived on the farm for eight years, he and Terces decided to get a couple of cows and begin milking them. Since then, Terces has become a master cheese, milk, yogurt, and cream maker. The cheese at dinner is indeed delicious.

The next day Matthew takes me on a tour of the farm. For self-admitted back-to-the-landers, the Engelharts are certainly industrious. In addition to their main house, they have constructed a large barn and numerous other farm structures. Rows of vegetables, movable green houses, and piles of compost all contribute to the cycle of life that is present here. It's a lot to have accomplished in eight years.

Be Love Farm farm grows vegetables, chickens, ducks, cows, and has a large fruit tree orchard. As we walk, Matthew hands me mulberries, peaches, and beans, all of which are brimming with flavor. He says, "Farms need to be much more diverse so that the soil microbes are more diverse, and fertility is cycling in a more powerful way."

Matthew tells me that their motto on Be Love Farm is "Annuals under the protection of trees, with the redemption of grasses." He explains that an annual begins life as a seed or a seedling "and it grows for a season, and it dies, and then the earth has to be plowed again. Whereas perennials, we don't disturb the ground." The grasses are important, he tells me, for moving carbon down into the soil, for attracting a variety of insects, and for feeding the cows.

Matthew explains that in the model of farming that they are practicing, they consider the soil "the internet." His objective, as much as possible, is not to disturb that network. "Every time we plow the ground, the soil internet goes down," he says. "The 'soil-food web' of microbes that are interconnected through mycorrhizal fungi, it really

BE LOVE FARM'S MODEL

"ANNUALS UNDER THE PROTECTION OF TREES, WITH THE REDEMPTION OF GRASSES"

On Be Love Farm, trees provide shade, deep roots, and leaf mulch. Grasses pump carbon down into the soil. Crops provide the food for their restaurant chain.

is an internet because it's a vibrant market, a bazaar, an exchange of minerals and energy. And so the key is minimum disturbance of the soil. And that's why perennial grasses are a big part of that; trees are a big part of that and no-till is a big part of that. And that's what we're experimenting with."

I ask Matthew why he is getting even more dedicated to farming now that he's pushing sixty, a time in life when most farmers are getting ready to retire. "We have two percent of the population growing our food. They're called farmers," he says. "And we've given them an impossible job because it would be like giving one teacher a thousand kindergarteners and expecting that we have nurtured, inspired students. We need more hands on the land. When you have more hands on the land, you can produce much more food per acre."

As we walk into a field of green plants, Matthew continues, "People say that organic agriculture or regenerative agriculture can't feed the world but the world actually is being fed predominantly by small farmers. There is way more food being produced by small farmers than large farmers."

It appears Matthew is correct. According to the Food and Agriculture Organization of the United Nations, 72 percent of farms worldwide are less than 1 hectare (2.47 acres). Twelve percent of farms globally are between 1 and 2 hectares. Especially in Asia, Africa, Europe, and Central and South America smaller farms grow the majority of food for the majority of people.[3]

But in North America the scenario is reversing, with farms getting larger while smaller farms disappear. This is why globally only 1 percent of all farms are larger than 50 hectares (123 acres) yet they control 65 percent of the land.[4] In the United States and Canada, large farms are typically dedicated to producing the commodity crops of corn, soy, hay, and wheat, which turn into animal feed and the types of carbohydrate-centric foods that go into a box, bag, or carton. This is why, in order for us to have a safe, sustainable, and sane food supply, Matthew says, "We need more farmers to produce more food, doing more intensive agriculture."

As utopic as the Engelharts' vision may sound, they are also pragmatic. Matthew believes firmly in a grassroots-based agrarian

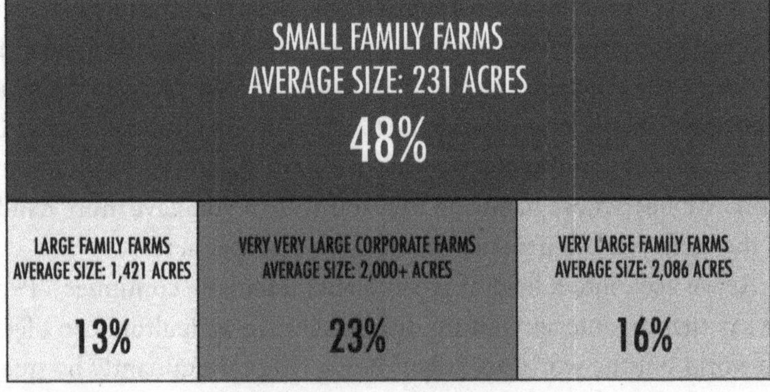

Today the vast majority of land farmed in America is still in the hands of small farmers, but an ever-larger amount of land is controlled by large farms and corporations. *Data:* USDA, various.

revolution, and he believes it can only happen from the cities outward. "Everyone can't be living on the land," he explains. "We need conscious city dwellers supporting and rallying for and championing more small farmers, and that will produce way more food and much better food." For Matthew and Terces, this is much more than a hobby; it is a mission.

The couple actually wrote the first book on "sacred commerce." After they coined the term, the idea took flight. Ayman Sawaf and Rowan Gabrielle, authors of the most recent book on the subject, *Sacred Commerce: The Rise of the Global Citizen*, define sacred commerce as "the party-cipation of the community in the exchange of products, information, and services that contribute to the revealing of the Divine (i.e., beauty, goodness, and truth) in all, and where spirituality—the return to the Self—is the bottom line."[5]

Sacred commerce is yet another iteration in the trend of business thinking loosely defined by terms like "the natural step," "conscious capitalism," and "the triple bottom line." Is it airy-fairy? Perhaps. But when Whole Foods founder and CEO John Mackey came out with his book *Conscious Capitalism: Liberating the Heroic Spirit of Business*, people listened. After all, this guy figured out how to grow a faltering health food store in Austin into a $14 billion-plus-a-year grocery store chain.

Call it whatever you want, this new type of business model is making money and creating jobs. It also seems particularly well suited to the food world.

RIPPLES

It turns out there wasn't a lot of competition for the Abounding River game I bid for on eBay on my smartphone at Café Gratitude. By the time I return home, I have long since forgotten about the auction. The game comes carefully packaged with a nice handwritten note in a little envelope. The note explains that the eBay seller overcharged me $3 for shipping. To compensate for the difference, she (I'm guessing it's a she by the handwriting) has included a book of stamps. I stare dumbfounded at the note. Who does such a thing?

The anonymous seller's note and her book of stamps are like an eddy—a small ripple in the river of my otherwise overscheduled, overcaffeinated, and underplanned day. It stops me in my tracks and makes me think about something Ryland told me. "Every day, with everyone who walks into our restaurants, our job is to 'underhand' them a little love, a little kindness, a little joy, a little transformation that can catalyze a ripple effect into the rest of their day, into the rest of their life. It's subtle, but subtleties make huge differences."

Eating an organic apple, or better yet, a regeneratively farmed apple, is different from eating a conventionally farmed one. Yet the difference is subtle. It's a little bit about flavor, a little bit about skin, a little bit about texture. Subtle things. The big thing is the process that brings that apple to the grocery store, farmers' market, or restaurant.

The lesson of Be Love Farm and Café Gratitude is that we foodies have the model reversed. It is not "farm to table" that we must focus on, but rather "table to farm." To Matthew and Ryland Engelhart's point, the foodie movement comes from the cities and dictates what is grown in the countryside.

Ryland summarizes it thusly: "I want my food to be paying a farmer a good, right livelihood. That farmer was saving water. That farmer was sequestering and pulling carbon out of the atmosphere, reversing global warming, and reversing ocean acidification. And he was feeding healthy communities so that they could continue to do that for generations to come."

Most people who eat in America today grow nothing. There are few who realize milk must come from a pregnant cow, even though upon inquiry this makes complete sense. Fewer still know that conventionally farmed food requires three pounds of toxic chemicals per American per year. And even fewer know that the process of growing organic produce requires the deaths of vast numbers of animals. Our choice for the future of food therefore is not vegan versus paleo versus omnivore versus vegetarian. Rather, we must choose between a food system that honors and respects the lives of flora, fauna, planet, and people versus a system that demoralizes, dehumanizes, and destroys our biological commons.

In our economy, demand dictates supply, not the other way around. If we are going to reform our food system to rebuild soil, to pull the carbon from the planet's oceans and atmosphere, and to feed 9.7 billion people, we must learn to substitute empathy for radicalism. We must understand there is no such thing as "no alteration" to nature and no such thing as "no death." Without exception in our biosphere, rhizosphere, and in the oceans, every living organism requires the death of another. Put in the simplest terms, death is a prerequisite for life.

But carnage is not a prerequisite. Nor is rampant suffering. Nor is the wholesale destruction of ecosystems and species. We no more need to acidify the oceans than desertify continents. Our species has a unique opportunity to relearn and truly understand our powerful role in the biological matrix of the planet on which we live.

To do this, we must relearn everything we think we know about what we eat.

We must begin with a clean slate.

TICK-TOCK

The next layer of industrial agriculture is already being developed and field-tested. It incorporates artificially intelligent computers, autonomous robots, self-flying drones, and advanced genetic designs. All this "intelligence" is based on the "more is better" principle of squeezing ever more nutrient-void calories from our depleted soils. In other words, more "overproduction."

Before we as a society sanction the next "green revolution," it is important to understand that there are alternatives to a food-producing machine that seems hell-bent on killing us. And some of those alternatives just might have the power to do other radical things like rebalance the Earth's climate and save our oceans.

If we are to survive on this little blue ball hurtling through space at 67,000 miles per hour, humanity must come to terms with this immutable reality: Our only possible future depends directly on our ability to save the very thing we stand on. Soil, dirt, sand, dust—call it what you will—the way we care for this precious foundation will determine the fate of our species. That fate is not a "someday maybe" kind of fate. It's today. It's right now.

Right. This. Second.

The clock is ticking. Time is in motion. But for all our button-clicking technology in this age of wonderment, we remain frozen.

Let this not be like a car crash seen in retrospect as though in slow motion. Let this instead be a defining moment for our abilities, our understanding, and our greatness.

Soil just might save us. But we are going to have to save it first.

I am not talking about a slow evolution of our food system. I am instead calling for an all-out revolution. Not a violent one. This revolution must be founded in the peace and respect and reverence for the very nature to which we are bound. It must be true to the Latin root of the word *revolution*: "revolutio," which means "a turn around."

Regardless of how you might feel about climate change the facts are simple. Carbon dioxide is acidifying our oceans. There, in the waters that cover the vast majority of our planet, is the basis for our oxygen and our food chain. If the oceans die, then we die. And while you or I may survive, our children's future is far less certain. And their children face an even more daunting tomorrow.

At the very least, a world with acidic oceans and toxic food is not something to strive for. Rather it is an apocalypse that we must not allow.

The time has come to stand together. The time has come for you and me to join the ranks. Slow food, permaculture, polycropping, biocybernetics, regenerative agriculture, organic agriculture, farm to table, the 4 for 1,000 program, vegetarianism, veganism, regenetarianism—these are all valid. It's time to find your place and sign up.

It's also possible to sign yourself up for radical change without becoming a radical. Food and eating food are our most powerful daily actions. When we interface with food, we make a choice that will determine the future. Thus, it is time to ensure that all the choices in the chain of events that lead to the act of eating are powerful as well.

It is time to peacefully, but firmly demand from our grocers, our restaurants, our farmers, and ourselves that our food be grown, cared for, processed, and eaten in accordance with the world we are now living in. We no more live in the post–Revolutionary War era of American westward expansion than the post–World War II era of suburban expansion. This is the twenty-first century, and we will soon have a planet of 9.7 billion humans. It is time we treat our food system, and each other, as such.

We know exactly how to feed ourselves and balance the planet's systems. We have the workforce. We do not need government assistance for this revolution to work. We need something much more powerful—individual and social commitment.

There are roughly twenty million unemployed young people in our country, and an estimated fifteen million Millennials living at or below the poverty line. This is the largest number of disempowered youth America has seen since the Great Depression. And yes, the conditions of that era are repeating themselves. Our Dust Bowl of today is more insidious, but as anyone who cares to go to the Midwest and California can see, it is there. As in the 1930s, we have a broken economy and a broken farming system.

Meanwhile, our farmers are retiring in droves. Ten thousand baby boomers retire every day. The majority of our nation's one million farmers with large farms will be retired within ten to fifteen years. Thus, there are two futures for America's farmland.

The first is that it will be gobbled up by the subsidiaries of megafood corporations. Like other vertically integrated industries, they will not directly own the means of production. They will "farm it out," so to speak, to third-party companies that can carry the liability of farming in a chemical war zone. This is the trajectory we are on today.

If we do nothing, your food, your children's food, and the food for generations hence will be controlled, from seed to table, by companies that produce calories and shareholder profits while starving humans of nutrition. This future is likely and it is predictable. It is not, however, inevitable.

Young men and women, now is the time for you to determine the future of your country and your planet. It is time to leave the coffee shops, the coding parlors of Silicon Valley, the wage-labor jobs of the fast-food industry, the debt-slavery institutions of higher learning, and the derelict and crumbing edifices of the dead urban sprawl of America. It is time to go back to the land.

Farming is hard work. Taking dead earth and restoring it to glory takes multiple years. It will not be easy. But what I am proposing is not an easy, overnight revolution. It is the kind of revolution where you dig in (literally) to make the kind of change that will last decades, even centuries.

To the farmers of today, it is time to choose. You have a limited time to learn, to mentor, to foster a better tomorrow. There will be great temptation to sell your land, to cash in and cash out. As sure as the sun sets, that payday awaits. But I assert that you have not given your life to the land for a check and a fifth wheel to take to Florida in the wintertime.

The farmers I meet have reverence for nature. They may not do buffalo dances in the light of the full moon, but there are few people in our modern society who have a better understanding of the cycles of nature than those who farm. Thus, my call to action to you is to put down the chemicals and sprays, those extensions of a Nazi experiment gone awry. Put away the tillage machines designed as a result of war. And find a young farmer, or better yet a whole group of them, to mentor.

For those of us who live in cities, our food is our power. Companies fear the buying power of a consumer base scorned. It is time for us to lead with our dollars and our choices. Some companies will ignore this disruption as an interruption while others will embrace it. Those who embrace it will succeed. Those who ignore it will vanish.

This is a deciding moment for humanity. Our fate depends on what we eat and how we produce that food. It depends on whether we continue to ignore our impact on the ecosystem or embrace the fact that we are inseparable from it.

The time has come for all of us to choose: Do we want a future rich with life or a future riddled with death?

If it is life we want, then the time has come for us to humble ourselves and kiss the ground.

CHAPTER 10

THE REGENERATIVE REVOLUTION

This chapter is a tool kit for restoring soil, revitalizing agriculture, balancing the Earth's climate, and giving your body the nutrition it needs.

Soil is the foundation of our civilization. If we regenerate it, we create the basis for healthy food for ourselves and for the generations to come. Agriculture drives how we treat soil. Thus, in order to regenerate soil, we need to build a new, safe, and sane food system.

To do this will require three components:

1. New federal, state, and local policies that eschew profit schemes for chemical and industrial agriculture and instead support the development of healthy soils
2. Hundreds of thousands, and potentially millions, of young people engaged in regenerative agriculture
3. And the most important component, a new, regenerative diet

This "regenerative revolution" aims to disrupt the status quo by combining these three components. The revolution I am speaking of is peaceful, open-sourced, and decentralized. To join this revolution, you don't need a badge, a pin, a membership card, or a secret handshake. There are no fees, no dues, and no rules of order. In fact, you can participate silently without telling anyone.

That's because, above all else, what we put in our mouths will drive the principles of regenerative agriculture on a local, national, or international level. Sequestering carbon dioxide into soil, reversing desertification, deacidifying our oceans and ensuring the survival of the oxygen-producing phytoplankton, creating a healthy food system that prioritizes a fresh and diverse diet over commodity bulk, incentivizing the humane treatment of animals—all these things are matters of our daily food choices.

Herein lies a watershed moment for our culture. The ultimate "diet" that America has been seeking for the better part of half a century is finally here. Enter: the regenerative diet.

But unlike a dieting fad, this is a lifestyle, a commitment not just to a better waistline but also to a better tomorrow. Yes, you can have the health and vitality you want, you can lose those extra pounds, and you can save the world just by a conscious, deliberate, and slow shift away from repackaged commodity bulk and toward the new plate outlined in chapter 9.

Americans spend at least $20 billion a year on dieting, and, not surprisingly, one in five Americans is on a diet right now. Most dieters will make four or five attempts to diet this year, they will fail, and then they will quit. The reason is simple: what we eat is a set of habitual choices. For most of us, food habits are formed early in childhood and reinforced by media, marketing, and social norms. Scientists tell us these habits are often ingrained on a cellular level.

The problem is that most diets try to stop you from doing whatever it is you are doing and instead rapidly do something entirely different. But this runs counter to how our brains are wired. Hence the immense financial success of the diet industry and the immense failure of dieters everywhere.

Unlike going on a diet, becoming a regenerative eater is a transition. This is a process that will likely take some time. It may even take the rest of your life. While its effects are indeed revolutionary, changing your dietary habits will only work if it is an evolutionary process.

You cannot think of this like a fad diet. This isn't a calorie-counting program, and it's not an attempt to rapidly change your deeply ingrained food habits.

This act of personal revolution is something totally different. It involves the "turning around" of the self. It is a deep and pervasive lifestyle shift brought on by a similarly complete mental shift.

It's also important to understand that to embark on a regenerative diet is to embrace a certain amount of failure. It is impossible to live in our society and eat a "perfect" regenerative diet. You will end up eating some things that are definitely not regenerative, certainly not organic, likely not good, and possibly even vile. I'm not saying you have to, but you likely will. And that's normal.

The idea here is a *general and continuing shift in what you eat and where it comes from,* not a "one-and-done," sharp left-hand turn away from your current food habits. As you shift what you eat, certain things that are now normal to eat will become abnormal. The key is not to try to abolish nonregenerative foods; it's to understand that they will always be there. It's just that, given practice and patience, your new plate of food can have less and less of them over time.

If you've flipped to the very end of this book to get straight to the "bottom line," here's the big takeaway: Now, more than ever before, you have the power to use your money, your voice, your hands, and, most important, your forks and knives to completely alter the future of the world in which we live.

The dilemma of what to eat is finally over. We can now get on with the business of building a food system that regenerates the nature on which it relies.

What follows is your personal "beginner's guide" to eating a regenerative diet, reversing climate change, saving the world, and becoming a regenerative revolutionary.

PERSONAL CHOICES

Food and Eating

1. **Reverse your plate: eat more veggies, less meat, and less processed food**

   In chapter 9, we looked at how chef Dan Barber is creating high-end food by "reversing" the typical plate to be veggie-centric rather than meat-centric. Root vegetables like carrots and sweet potatoes along with protein-centric beans like fava and chickpea offer a lot of opportunity to make the "center cut" of your family's dinner the veggies. Good news—it's cheaper than meat, even if you buy these vegetables as organic. This also doesn't mean forgoing meat entirely. The key is to see how meat can be 25 percent or less of your total calories while the other 75 percent is whole grains, fresh vegetables, organic dairy, fruits, and foods that do not come from a bag, box, or carton.

2. **Start with lunch**

   For many people with families who work, breakfast and dinner can be chaotic. But lunch is sometimes the only real break we get during the hectic day. Use lunch as an opportunity to reverse your plate. Instead of eating a sandwich consider using a bed of greens and adding your favorite items to it in a big bowl, throwing in some salad dressing, and voilà!—your bread and meat-centric lunch has been replaced with something that won't make you feel bloated and sleepy. To make it easier, greens can be purchased prewashed in boxes. Other items, like nuts, seeds, deli meat, et cetera, can also be bought prepackaged. The goal here is not perfection, but rather an improvement. Your body will thank you for salad-centric lunches, too!

BEST PLANT-BASED SOURCES OF PROTEIN

VEGETABLES

FOOD	SERVING SIZE	CALORIES (CAL)	PROTEIN (G)	% OF CALORIES FROM PROTEIN
Spinach, cooked	1 CUP	41	5	49%
Asparagus	1 CUP	27	3	44%
Broccoli	1 CUP	31	2.6	34%
Brussels sprouts	1 CUP	38	3	32%
Peas	1 CUP	118	8	27%
Sun-dried tomatoes	1 CUP	139	8	23%

PROTEIN POWDER

FOOD	SERVING SIZE	CALORIES (CAL)	PROTEIN (G)	% OF CALORIES FROM PROTEIN
Pea protein powder	1 OUNCE	103	24	93%
Brown rice protein powder	1 OUNCE	110	15	55%
Hemp protein powder	1 OUNCE	113	13	46%

NUTS AND SEEDS

FOOD	SERVING SIZE	CALORIES (CAL)	PROTEIN (G)	% OF CALORIES FROM PROTEIN
Hemp seeds	1 OUNCE	162	10	25%
Peanuts without shells (technically a legume)	1 OUNCE	164	7	17%
Black walnuts	1 OUNCE	173	7	16%
Flax seeds	1 OUNCE	110	3.8	14%
Chia seeds	1 OUNCE	138	4.7	14%

BREAD, GRAINS, PASTA

FOOD	SERVING SIZE	CALORIES (CAL)	PROTEIN (G)	% OF CALORIES FROM PROTEIN
Seitan	1/2 CUP	180	31.5	70%
Oat bran, cooked	1/2 CUP	44	3.5	32%
Whole wheat pasta, cooked	1/2 CUP	87	3.5	16%
Buckwheat flour	1/2 CUP	291.5	11.5	16%
Wheat flour	1/2 CUP	203.5	8	16%
Whole wheat bread	1 OUNCE SLICE	77	2.9	15%
Quinoa, cooked	1/2 CUP	111	4	14%
Oats, cooked	1/2 CUP	153.5	5.5	14%

BEANS AND LEGUMES

FOOD (ALL BEANS ARE COOKED)	SERVING SIZE	CALORIES (CAL)	PROTEIN (G)	% OF CALORIES FROM PROTEIN
Tempeh	1/2 CUP	160	15.5	39%
Soy beans	1/2 CUP	127	11	35%
Brown lentils	1/2 CUP	115	9	31%
Red lentils	1/2 CUP	115	9	31%
Green lentils	1/2 CUP	115	9	31%
Kidney beans	1/2 CUP	109.5	8	29%
Split peas	1/2 CUP	115.5	8	28%
Lima beans	1/2 CUP	108.5	7.5	28%
Black beans	1/2 CUP	113.5	7.5	26%
Black-eyed peas	1/2 CUP	99	6.5	26%
Tofu	1/2 CUP	94	6	26%
Pinto beans	1/2 CUP	122.5	7.5	24%
Navy beans	1/2 CUP	127.5	7.5	24%

Source: Matt Frazier, www.NoMeatAthlete.com.

3. Begin to eat a regenerative diet

A regenerative diet is a powerful new tool in combating climate change and eating well. The diet can be vegan, vegetarian, or omnivore. Essentially, it involves choosing products at the store, the farmers' market, and at restaurants that have been grown in a regenerative way. This will involve asking how and where the food was grown. It means making sure your veggies and grains come from farmers using very little tillage and practicing cover cropping. It also means passing over the traditional meats in favor of meats from animals raised humanely, on pasture, and killed with the least amount of suffering possible.

4. Clean out your cabinets, pantry, freezer, and fridge

You can't make new choices with all the old food hanging around. Remember, 63 percent of the calories in an average American diet come from things like fats, sugars, and processed junk that is mostly derived from chemical-laden corn, soy, and wheat. A simple guide to start: If the item in question has processed sugar, such as cane syrup or just "sugar," get rid of it. The vast majority of foods that have sugar as a main ingredient are repackaged commodity crops. The first time you do this "cleanse" it may be daunting. Do not fear! There are many other nutritious things that will soon fill your kitchen.

5. Make your refrigerator a display of nutrition

There's almost nothing more interesting (and scary) than looking in other people's fridges. Most fridges are a mess. They are overstuffed with things that barely classify as food. Often, the items most visible and accessible are the worst for you. If you are the foodie in your household, take on the job of making your fridge a presentation. The items that are front and center can be healthy and nutritious grab-and-go things: fruits, vegetables, whole grain breads, salad items, et cetera. Make sure condiments and add-ons stay on the door shelves, not on the main shelves.

Keeping your fridge clean, tidy, and presentation-ready takes a bit of work, but it is one of the most important tricks to changing your and your family's diet. *Photo:* Dalia Taher, health coach, www.qijuices.com.

6. Do you have a fruit bowl in the kitchen?

Like most animals, humans tend to eat what is most visible, most accessible, and easiest to grab. A household that has a large bowl filled with seasonal fruit will be a healthier one and will want less for items from a box, bag, or carton. Fill your bowl with apples, oranges, grapes, bananas, or anything that can easily be grabbed and eaten on the go. Put a Post-it or a sign on the bowl: EAT APPLES! Remember, this is about you and your family transitioning, not about perfection. It may take a few tries, and some prompting, to get people to leave the Oreos for the organic apples.

7. Do a one-week food accounting

The number one complaint people have about eating an organic, healthy diet is that it is "too expensive." But over half of what America eats is reprocessed bulk calories that have a hefty markup. In other words, to transition to a more regenerative diet, you need to be honest with yourself about where your food is coming from and where your money is going right now. To do that, do a one-week accounting of all the food you buy and how much money you spend. Once you truly understand what you spend your money on (all those Doritos and Cokes really do add up), you can begin to honestly compare how much those "expensive" items like organic bread, organic eggs, and the like cost. There are some food journaling apps as well as some expense tracking apps that might be helpful for this.

8. **Change your kitchen sweeteners**

Most kitchens have at least one bag of grain sugar, at least one bottle of fake maple syrup (which is actually corn syrup), and often a bag of confectioner's sugar. Try substituting these items for stevia drops, agave syrup, actual maple syrup, and raw, unprocessed honey. The more you use low-glycemic sweeteners, the more your palate will adjust. Remember, most of the drive to eat items in a bag, box, or carton comes from the desire to taste sugar. Getting your kitchen, your family, and your own mouth off sugar will save you money, it will save your body, and it will open your shelves and your life to more nutritious and regenerative food choices.

9. **Create a food cheat sheet and post it somewhere visible**

There are simple rules for being vegan, vegetarian, paleo, and macrobiotic but when it comes to being a "regenerative eater" it's not so cut and dried. This is why our family created our own Food Cheat Sheet, which we laminated and posted in our pantry where much of our food is stored. Your cheat sheet should give your family ideas for food and snacks that are fun, easy, and inexpensive. For your reference, I have included our cheat sheet, but by all means make your own!

10. **Make a place to advertise your "weekly specials"**

I do our family's food shopping on Sundays, when our town has its farmers' market. Once I'm done at the market, I'll hit the health food store and grocery store for items we're still missing. Once I'm home I write our "Weekly Specials" on a chalkboard hanging in our kitchen. Sometimes this is something that's in season like "Blood Oranges—Mmmmm" and sometimes it's something that was on sale. The point is that when you begin to purchase first from

OUR GOOD FOOD & GOOD HABITS SHEET

STEP #1 – LOVE YOURSELF, LOVE YOUR BODY, LOVE EACH OTHER

Nobody is perfect. Health is a journey, not a destination.
Forgive, love, and rejoice every day. Take time for self-care.

STEP #2 – BURN, BABY, BURN!

To lose 1 pound per week, burn 500 more calories than you eat daily.

Calories Burned	Activity
200 calories	60 min. fast walking
300 calories	60 min. of yoga
400 calories	60 min. sweat workout
600 calories	60 min. running

STEP #3 – AVOID FOODS THAT PRETEND TO BE YOUR FRIEND

Avoid: fried foods, milk & dairy, bread, bagels & chips, pastas, nonorganic soy, and these:

Calories Added	Food
350 calories	Sugar & Sweeteners (average daily)
110 calories	Cheese (1 slice)
100 calories	Bread (1 slice)
160 calories	White potato (1)
150 calories	Beer or Wine (1)
250 calories	Rice (1 cup) (also pasta / cornmeal)
120 calories	Butter, fat, oil (1 tablespoon)
500 calories	French Fries (1 serving)
300 calories	Tortilla chips (1 cup)
400 calories	Cake (1 slice)
500 calories	Coconut milk (1 cup)

STEP #4 – EMBRACE THESE FRIENDLY FOODS

An active woman in her late 30s needs approx. 2,000 calories a day; an active 40-year-old man needs approx. 2,500 calories a day. Read labels and get your calories here:

MEATS: salmon, tuna, organic baked (not fried) chicken, lean grass-fed beef (no pork)
FATS: avocadoes, small amounts of goat cheese, olive oil, peanuts
PROTEINS: eggs, almonds, walnuts
FRUITS: grapefruit, bananas, apples, oranges, berries
STARCHES: black, kidney, and garbanzo beans (no pintos, no soy), lentils, quinoa, squash
CAFFIENE: green tea (preferable to coffee)
TREATS: dark low-sugar nonmilk chocolate, dates, dried fruits, raisins (organic)
GREENS: broccoli, cauliflower, cabbage, salad (careful of dressing—it's all fat)
MUNCHIES: carrot sticks, celery sticks, humus, guacamole, non-buttered popcorn

farmers, and secondly from the seasonal fruits and veggies in the grocery store, the primary food items in your house will shift with the seasons. Whether it's a dry-erase board stuck to the fridge, a chalkboard, or some other sign, find a way to let your family know what's good to eat this week.

11. **Check your fish**

It's no secret that the world's oceans are in trouble. Overfishing has caused a long list of species to go extinct or nearly extinct. Obviously, pressure must continue to be applied against whaling and dolphin killing operations in Asia until such practices are abolished. But your personal fish-eating choices matter too. Seafood Watch publishes a wallet-sized booklet, has a website, and an app that you can check before ordering that fillet. Download the app through Seafoodwatch.org.

12. **Be a vegetarian for a month, a week, a day, or even a meal**

For those of us who have grown up eating meat three times a day, becoming a vegetarian might seem like a radical move. But you don't have to be a radical to give your body a break. Afraid? Just go to a reasonably normal restaurant and ask the waiter about their vegetarian options.

13. **Compost your food scraps**

Nature composts everything. While you can't compost all your trash you can compost most of your table scraps. There are two basic forms of compost—without worms and with (vermicompost). There are compost tumblers and containers for homes of just about any size. There are even vermicompost rigs you can use in your apartment. The soil produced is usually rich and dark and can be used on anything from potted plants to your garden.

14. **Work with your children to pack their lunch**

I get it, life is busy and hectic and who has time to pack their kids' lunches? One trip to your children's school cafeteria and you might change your mind. It should be no surprise that prison food and public school cafeteria food are supplied by a lot of the same vendors. Ask your children what they would like to see in their lunch box or bag. Take them to the farmers' market and the grocery store and pick out items together you know they will eat. Or simply go on Pinterest, where there are a thousand-plus photos of healthy school lunches. Make suggestions and compromises. Don't have time to pack lunch in the morning? Do it the night before. Put the items you need for their lunches on a cabinet or refrigerator shelf so you can prep quickly. Try water with lemon juice (or herbal tea) and stevia instead of sugary drinks. Your kids will be healthier and more alert at school, and, who knows, they might even start to educate other kids.

15. **Remember that all food is sacred and to practice forgiveness and gratitude**

It's easy for those of us who live in a world of relative food abundance to judge ourselves and others for what we eat. But about one in nine people does not have enough food. When you eat, which in America is statistically more often than most people, it's important to be thankful for every bite. It's also important to forgive yourself for the food choices that may not be "perfect." Left unchecked, "food guilt" can turn into a psychological disorder. Even if you don't believe in a higher deity, there are a number of studies that actually show that taking time to bless your food can have a positive effect on your body and may, in some cases, even have a positive effect on your food as well. Amen.

The Power of Your Purchase

16. **Buy organic**

Money speaks loudly to the companies that provide the foods we eat. The USDA Organic label is considered by people in the regenerative agriculture movement to be the "baseline standard"; in other words, the minimal accepted standard by which food should be grown. Want to move the agribusiness machine away from chemicals and GMOs and toward practices that will sustain our species for generations to come? Well, it begins with buying products that have this label. Look for it on your fruits, vegetables, milk, chickens, eggs, and meats. If it's not there, ask for it. Eventually, there will likely be some kind of "transitional regenerative" logo as well as a "regenerative organic" logo to look for, but these may take some time.

When possible, look for the USDA Certified Organic Label on your food, and make this the "baseline" for what you buy.

17. Look for these labels

USDA Organic is a great start, but especially in the case of animal welfare and health, we need stricter and more holistic government standards. In the meantime, there are a number of groups that provide a guide for foods that are raised with far more care than the industrialized, post–World War II model of factory farming. Look for these labels:

American Grassfed
American Humane Certified
Certified Humane
Animal Welfare Review Certified ("Animal Welfare Approved")
Global Animal Partnership
Certified Sustainable Seafood MSC (Marine Stewardship Council)
Biodynamic
Bird Friendly

18. Shop at your local farmers' market

Most cities and towns now in the United States have a weekly farmers' market. It's time to get yourself there on a weekly basis. You can usually get many of the basic products you need, including salad makings, fruit, vegetables, eggs, meat, and dairy products, from people about whose products and practices you can ask directly. Don't see something you want or need? Ask. The communication between producer and consumer is one of the cornerstones of regenerative agriculture.

19. Demand transparency at the grocery store and restaurants

Let's face it, most restaurants are "don't ask, don't tell"–type places. The attitude is usually that "all that stuff" (food selection) is done by "professionals." For food choices while eating out to matter, restaurateurs are going to have to feel the pressure from customers to make choices that have to do with the health of the land and the foods they choose. Even if it says "farm to table," ask what that means. What farm did it come from? It's pretty easy to check if the advertisement matches the reality. For these terms to have meaning, consumers are going to have to get much more active about ensuring they aren't just labels.

20. Support restaurants that serve organic, farm to table, and especially regenetarian cuisine

The local eatery with USDA Organic foods may not have your favorite whatever (chair, table, dish, et cetera), but if one of these types of places exists in your area, it's time to vote for its continued success with your money. Remember, when you pay for this type of food, your money is paying for an entire chain of events that begins with how the soil is treated and often goes all the way past your plate into the compost that will be made from your scraps. If you really need that chair, table, dish, et cetera, then speak to the owner or manager and tell them the presence of this thing would make the difference of your patronage or not. They may be more than happy to provide it.

21. Join a CSA

There are thousands of community-supported agriculture farms and gardens across the country and around the world. CSAs are membership driven, and members receive a basket, bag, or box of what is being grown right now. Some offer delivery but most of them have established pickup times for members to come and get their share of fresh produce on a weekly, bimonthly, or monthly basis. LocalHarvest.org has maintained an active CSA directory for many years. A great book on this subject is *Sharing the Harvest: A Citizen's Guide to Community Supported Agriculture.*

22. Shop around the edge of the supermarket

While no guarantee for nonchemical food, the edges of the supermarket (as opposed to the middle) are your best hope for finding fresh produce and organic products. Venture into the middle of the supermarket and chances are you'll be looking at something in a bag or box made of processed corn, wheat, or soy.

23. Get your children's school to offer healthy, organic foods for lunch

Americans pay taxes so that our children can have a good education. Nowhere is it written in stone that they should be fed toxic food in the process. But go to an average public school cafeteria and you'll be hard pressed to find a food that hasn't come from a box, bag, or carton from a mega-food company. With foods filled with sugar, salt, and chemicals it's no wonder our students are falling behind the rest of the world. The Good Earth Natural Foods program in Northern California is an example of what is possible in terms of school lunches. Serving 1,300 organic school lunches each day, this public-private partnership shows what happens when a local business takes an interest in the future of its community. Your school can't afford better food? Maybe it's time for a local company or consortium of companies to step up and pay for it.

24. Invest in regenerative agriculture programs

This is still a new area, but there are more and more socially responsible funds. Called impact investing, this is an area that is quickly growing. Keep an eye on this, as there are sure to be more opportunities for investors to support regenerative agriculture projects. Alternatively, you can help a farmer get started by investing in a local regenerative agriculture farm.

25. Buy clothing from regenerative sources or from thrift stores

The Fibershed project aims to produce clothing locally from sources that are regeneratively grown. This may sound ambitious but our clothing takes tremendous energy and water resources to produce and the process of creating textiles usually involves substantial amounts of chemicals. There's also a labor-rights issue, with most clothing being sourced from low-wage countries. For more info, see www.Fibershed.com. Can't afford regeneratively grown clothing? No worries. Thrift-store shopping is in a very real sense recycling (and it's fun).

26. Support hemp products

Hemp is not a perfect crop, but it has numerous benefits over the current industrial crops. From hemp clothing to hemp foods to hemp shoes, this one plant offers many products for humanity. Hemp foods contain healthy omega-3s and are a good source of plant-based protein. The hemp plant can be turned into clothing, rope, plastic, and other industrial products. Be sure to support USDA Organic hemp because as good as hemp may be, it is also being grown today with heavy herbicides and pesticides.

27. Join a carbon offset program

Carbon offset programs come in many flavors, including offset programs for individuals and businesses. The basic idea is that you are "purchasing" a method of carbon dioxide reduction. This is generally accomplished by planting trees. Check the program carefully before signing up, especially the integrity of their operations.

Education

28. Teach your children about food, agriculture, and ecological systems

It's easy to stay inside and let the TV and our gadgets become the center of entertainment. Try turning off the screens for an hour and go to a local park, on a hike, to the mountains or the beach. Fourteen percent of the landmass of the United States is protected federal land, including state parks. Go outside; take your kids and yourself for a walk. Talk about nature. Engage them in an understanding and awareness that they are part of this magical planet we all share.

29. Attend the National Heirloom Exposition (and others like it)

Perhaps the best event in the United States around getting back to the roots of our food is the National Heirloom Exposition in Santa Rosa, California. Generally occurring in early September, this event offers education and fun for children and parents alike. From the speakers to the food to the vendors and presentations on seeds, plants, and animals, this is by far one of the most engaging events connecting people to their food.

Farming and Gardening

30. Volunteer on a farm, intern on a farm, or join WWOOF

I've never met a farmer who wouldn't accept free labor. Working in agriculture, be it for a weekend, a month, or a year, will give you a lasting and powerful connection to your food. It will give you something else, too—power. When you know how to grow and manage your own food you are no longer dependent on companies and governments to do that for you. This too will change your perspective. There are a number of farmer internship programs. I recommend WWOOF—World Wide Opportunities on Organic Farms. Their network is global and their organization has been around for a long time. Become a WWOOFer and you will not regret it.

31. Get your lawn off chemicals

Lawns are the most heavily sprayed and irrigated crop in the country. It used to be that having clover in your lawn was a sign of health. Today, for most homeowners, it is cause for a Vietnam War–like herbicide-spraying campaign. Let's face it—lawns today have gotten ridiculous. They are sprayed with ten times more herbicides and pesticides per acre than our crops. The most common herbicide? 2,4-D. These poisons are in direct contact with our children, our pets, our parents, and ourselves. If you have a lawn, manage one, have parents who have a lawn, or know somebody who does, get it off chemicals. Get your soil tested or do your own pH test: Lawns need to be around 6.5 or 7 on the pH scale. Find some locally adapted grasses. Like Gabe Brown in chapter 8, think of your lawn like a "cover crop mix" of clover, grasses, and small legumes. Don't cut the lawn too short. Most turf grass needs to be at least three inches to sustain healthy foliage and roots. With some experimentation and work, your lawn can become a carbon sink, and you, dear lawn owner, can become a regenerative farmer.

32. Start your own food garden or planter boxes

Many cities do not own that strip of turf between the sidewalk and the street. That's why in cities across the country, nicely made planter boxes are popping up next to sidewalks. In a foot or two of soil you can grow a tremendous amount of food, including lettuces, carrots, potatoes, and even tomatoes. Wouldn't our cities be more beautiful with thousands of species of food growing between our streets and homes? Planter boxes also work on apartment balconies. Even better—have a little spot of bare earth, asphalt, or cement just sitting there? Get a pickax, get rid of that unused hard surface, and plant some food. There are literally hundreds of books and websites on planting your own small-scale garden and building planter boxes.

33. Get connected to your farmers and visit them

Farm tours are fun, and often all you have to do to be invited to your local farm(s) is ask. Talk to the people behind the stands at your local farmers' market. Where is their farm or ranch operation? Do they ever do tours? Can you bring your kids? Most farmers love showing off their operation. If they don't want to show their farm off, that should be a red flag. By getting to know your local farmer, you create a connection with one of the most important people in your life. Farm tours are also popping up as events across America so take advantage of those when they happen.

34. Save seeds

For centuries humans saved the seeds of their best crops and plants to pass on to the next generation. You can save the seeds of any plant you appreciate. Seeds can be saved in simple mason jars (or cleaned jars from store-bought products). By saving seeds, you help preserve genetic diversity, save money, and can participate in really homing in on the types of plants that work together in your growing environment. This is also a fun activity to do with your children (or parents). Again, there are a lot of great books and online resources on this topic.

35. Start your own beehive

Due to the neonicotinoids found in pesticides and other environmental factors, bees have had a massive population collapse. Bees are critical to the ecosystem and they are critical for humans, too. Starting your own beehive can be both rewarding and helpful to your local ecosystem. First make sure you are not allergic to bee stings. Check local beekeeping regulations, join a beekeeping club, get gentle bees, make sure if you're in a suburban neighborhood that your fences are wood and at least eight feet tall (so the bees will fly over people's heads), speak to your neighbors, and "be cool" by letting the bees do their thing and not disturbing them too much.

36. Plant trees (or just one tree)

There's perhaps no single planting action that has as many benefits as putting a tree in the ground. Trees sequester carbon dioxide, they help maintain the water cycle, they cool the localized air, they provide shade, they stabilize the soil, and they are pleasing to look at. There are so many cool organizations committed to planting trees in different parts of the world that need funding and volunteers, so get involved.

37. Read *The Carbon Farming Solution*

If you are a farmer or want to become one, I highly recommend reading *The Carbon Farming Solution* by Eric Toensmeier. It is a pricey textbook-like tome but every page is revelatory and what you spend on the book you'll save many times over while farming.

INITIATIVES

1. Get lawns, parks, golf courses, and gardens off chemicals

Your child is playing in it. Your dog is rolling in it. Your spouse is spending hours golfing on it. I'm talking, of course, about 2,4-D, the widely sprayed herbicide that was part of Agent Orange. Commercial lawns, parks, golf courses, and gardens are doused in far more chemicals than our food, and then we play on that ground. It's time to involve city and state governments as well as lawn and landscaping companies in soil education. Join the movement to ban toxic sprays on all land that the public interacts with. This movement is happening from Marin to Minneapolis but at the time of writing there is not yet a hub. A good rule of thumb is to make sure your park, golf course, et cetera, becomes USDA Organic.

2. Establish school gardens and organic lunches

Many schools have unused playground areas that can be converted to gardens. Combined with a program to deliver healthy food for lunch, school gardens can be a critical learning tool for sciences, including biology and ecology. There is a growing online resource base of support for establishing school gardens, including grants, examples, and nonprofits that are committed to helping get a garden into your child's school. At the time of writing Whole Foods even offers financial assistance through their Whole Kids Foundation program.

3. **Start an agroecology charter high school**

There are over six thousand charter schools in the United States and that number is growing. Charter schools offer parents and teachers a way to construct a new education experience as opposed to one that is entirely focused on standardized testing, like so many public schools today. Just ask the team that created Golden Bridges School in San Francisco. Their modern living buildings will be the first urban farming grade school in America. Don't forget, the French government has a complete educational curriculum for high schools that focuses on soil and agroecology. Using the French model as a template American charter schools can get a leg up on adopting these educational methods.

4. **Start a city compost program**

In Sacramento, a group called BioCycle wanted to take the city's composting program to the next level. They designed a custom-made bicycle trailer that could pick up thirty-two-gallon barrels of waste. The BioCycle team now picks up compostable materials from restaurants and drops off finished compost to gardens. The compost programs of many cities were started by active citizens committed to seeing change. Do your research, print out examples, and let program directors know that compost can be an income source.

5. **Bolster the organic standard and call for better USDA supervision**

The USDA Certified Organic Standard is the only government program in America that assures food is not sprayed with toxic chemicals. The standard calls for increasing organic matter in soil as well as many other techniques that would be soil beneficial. But the spirit of organic as well as the on the ground regulation of farmers are often not enforced. For the reality of USDA Organic to match the ideals set forth on paper, more and better regulation is needed as well as more and better education of farmers who are transitioning to USDA Organic. Encouragement and pressure are needed toward the Department of Agriculture to make organic better.

6. Help End CAFOs forever

Birke Baehr was just eleven years old when he gave a TEDx talk entitled "What's Wrong with Our Food System?" The talk, which as of the time of writing has over 2 million views, was powerful because it showed a young person speaking about issues including concentrated animal feeding operations (CAFOs). If an eleven-year-old can reach over 2 million people with his message of ending CAFOs, what can you do? There are many nonprofits, petitions, and consortiums dedicated to ending this practice, which poses health risks to both humans as well as animals.

7. Start or join an urban garden

After a long hiatus after World War II, urban gardening is making a comeback. People are Instagraming, Pinteresting, and generally sharing about the urban gardening phenomenon online like never before. The derelict metropolis of Detroit is well known for its inner city urban garden programs that turn blight into greenery. Get on the bandwagon and turn that old parking lot or unused lot into a food-producing plot.

8. Educate women in developing countries

One of the most important ways to bring down carbon emissions is by educating young women in the developing world. Women with educations tend to have fewer children, burn less firewood, and generally move their communities in a more positive, stable, and sustainable direction.

9. Start a seed bank

The 1920s-era bank building in downtown Petaluma has a new purpose: storing seeds. A "seed bank" is a place where people can go to purchase seeds, learn about cultivation, and get their "garden" on. Seed banks are great community centers and help foster local gardening and farming.

10. Start or join a rooftop garden and get your city to do green roofs

Building rooftops in suburban and urban areas can be great locations for community gardens. Done in large enough numbers, they can help fight the "heat island" effect of cities and provide a much-needed natural oasis. Chicago is known for its vast expanse of green roofs. There is even a Green Roof Professional Training Program (GRP). Check out Greenroofs.org for more information and resources.

11. Start an inner-city program to deliver fresh produce and meats

Many inner cities are what are known as "food deserts"—locations that are far from a grocery store that sells fresh produce. These deserts have a disproportionately negative impact on people lower on the socioeconomic ladder and contribute to the food insecurity that many low-income people face. Enter programs like Oakland's People's Grocery, with their Mobile Market truck that sells fresh food to members in its area. The grocery truck also accepts food stamps. After successfully providing a mobile grocery service for over a decade, the People's Grocery inspired a new nonprofit called People's Community Market, which is working to build a fourteen-thousand-square-foot neighborhood grocery store, health hub, and events center.

12. Get EBT/food stamps accepted at farmers' markets

There are 46.5 million food stamp recipients in America (almost 23 million households). EBT stands for Electronic Benefit Transfer and it's the name given to today's debit card–like food stamp program. Starting in 2016 Los Angeles voted that all the vendors at its farmers' markets must accept EBT cards. This is a positive for the farmers as they still get paid for their produce. And it's good for EBT holders too, as it gives them access to reasonably priced healthy food they otherwise might not be able to get. Make sure your local city, municipal district, and state have laws that enable EBT use at farmers' markets.

13. **Get insurers to cover the transition period for a farmer from conventional to organic**

During the required three-year transition period to become a USDA Certified Organic farm, a farmer will often not qualify for crop insurance under either the traditional crop insurance program or the organic program. This adds tremendous financial risk to the already difficult process of going organic. Insurance programs must be created to help farmers move from chemical dependence to independence. This could be a financially rewarding program for all involved.

14. **Start a National FFA Organization chapter at your high school**

The National FFA Organization (formerly Future Farmers of America) is a chapter-based, government-supported program that can be started in any school in the country. Currently there are around 7,700 chapters with almost 630,000 members nationwide. FFA programs can provide education, technical assistance, and grants to students who want to get involved in farming and ranching. Much of the education is around "traditional" industrial agriculture, but the program is open-source enough that students can help mold their own chapter's goals. There is even a government grant for starting a chapter at your school! See FFA.org and USDA.gov for resources.

15. **Start a Kiss the Ground chapter at your university**

Using the model developed by the Public Interest Research Group (PIRG), university and college students can start a chapter of Kiss the Ground to get involved in agroecology, agroforestry, and regenerative agriculture in their community and school. For more information, see www.Kisstheground.com.

16. Become a farmer or rancher—get government money

There are numerous loans for first-time farmers available through the USDA. The Farm Service Agency (FSA), a branch of the USDA, offers a number of low-interest loan programs, including a "microloan" of up to $50,000 to help with farmland and building purchases as well as other farm improvements. Their loans for "Beginning Farmers and Ranchers" can be up to $300,000, and their loans for folks wanting to start farming can be over $1 million. They even have loans for women farmers and farmers from minority ethnic groups.[1]

17. If you're a farmer, become a mentor

There are a number of regional farmer-to-farmer mentorship programs and a couple of national programs. Midwest Organic and Sustainable Education Service (MOSES) is one such organization that pairs beginning farmers with experienced ones in the organic space. If you are a farmer and want to offer mentorship to young people starting out, check out your local farmer mentorship programs as well as the local chapters of the FFA.

18. Farmers—use the COMET-Planner to get into regenerative agriculture

The NRCS and Colorado State University have created an online tool that helps farmers calculate their carbon dioxide emissions or sequestration. The tool is called COMET-Planner and is located here: www.comet-planner.com/. The site also contains information on conservation planning techniques that the NRCS uses to create its regenerative agriculture plans.

19. Start a reforestry program in your area

There are very few towns and cities that could not do with more trees. Added to this is the loss of topsoil every day. With the constant pressure of desertification affecting over 1 billion acres globally, there are endless opportunities to plant trees. Numerous nonprofits are available as resources to help guide you in the process of creating a local reforestry program to help your area.

20. Sign on to the 4 for 1,000 program

As a component of the United Nations Paris-Lima Accord, Minister Le Foll's program is available to states, local authorities, companies, farmers' organizations, NGOs, and research institutes. It's free to join. Go to http://4p1000.org/understand.

21. Lobby to rewrite laws around commodity crop insurance

Research groups such as the Robert Wood Johnson Foundation in Princeton, New Jersey, recommend leveling the playing field by extending insurance programs more widely to fruit and vegetable producers. The Center for Rural Affairs agrees, saying crop insurance should be "crop neutral." The imperative is clear: Any new farm policy should, at the very least, remove the current perverse incentives for people to eat unhealthily. We need policies that support soil rather than commodity agriculture.

22. Lobby for the Environmental Protection Agency (EPA), the Department of Environmental Quality (DEQ), and the US Department of Agriculture (USDA) to test for glyphosate, 2,4-D, atrazine, and the five most commonly used commodity crop herbicides and pesticides

Beyond Pesticides is an organization dedicated to ending the use of toxic chemicals on our food. Their blog and petitions offer opportunities to get involved in the effort to end the use of these substances forever. Go to www.beyondpesticides.org.

SUPPORT AND ADVOCACY ORGANIZATIONS

4 for 1,000 Program (http://4p1000.org/understand)
International Federation of Organic Agriculture Movements (IFOAM) (https://www.ifoam.bio)
Regeneration International (http://regenerationinternational.org)
Rodale Institute (http://rodaleinstitute.org)
Women, Food, and Ag Network (https://www.wfan.org)
Quivira Coalition (http://quiviracoalition.org)
Project Drawdown (http://www.drawdown.org)
The Carbon Underground (https://thecarbonunderground.org)
Beyond Pesticides (http://www.beyondpesticides.org)
Center for Food Safety (http://www.centerforfoodsafety.org)
Environmental Education Media Project (http://eempc.org)
Kiss the Ground (https://www.kisstheground.com)
NRCS (http://www.nrcs.usda.gov)
Fibershed (http://www.fibershed.com)
WWOOF (http://wwoofinternational.org)
Local Harvest (http://www.localharvest.org)
Soil Not Oil (http://soilnotoilcoalition.org)
Moms Across America (http://www.momsacrossamerica.com)

ACKNOWLEDGMENTS

My deepest gratitude and loving thanks to my wife, Rebecca, for her belief in me and in this book. During my darkest hours, your love, enthusiasm, and support never wavered. This work is an expression of our partnership and our purpose. A heartfelt thanks to my daughter Athena, who at a young age was supportive of her father all the way through.

Thank you to my publisher, Zhena Muzyka, at Enliven Books, without whose incredible tenacity and unbeatable positivity these words would still be sitting on a hard drive. Thank you to Judith Curr at Atria Books and the entire staff at Simon & Schuster, especially Haley Weaver and Albert Tang. Thank you to Miles Kelly, who created the unique graphics in this book.

Thank you to Minister of Agriculture Stéphane Le Foll and his staff. Thank you to Anne Moyat, who single-handedly wrangled the French government into agreeing to let an American journalist into their cloistered world in the middle of a terrorism crisis. I owe the people of France great gratitude for their openness during such a vulnerable time.

Thank you to Ryland Engelhart, John Roulac, and Darius Fisher, who were each instrumental in inspiring this work. Thank you to Kristin Ohlson, whose critically important book *The Soil Will Save Us* laid the groundwork for my own journey. I also want to extend a special thanks to the many farmers and ranchers who have hosted me over the past two decades. Thank you to Doniga and Erik Markegard for letting me into your lives. Thank you also to the many people who woke up early, stayed up late, went out of their way, and patiently gave me their time, including Robert Reed, Gabe Brown, Ray Archuleta, Tony Ten Fingers, Paul Hawken, Matthew and Terces Engelhart, Rebecca Burgess, John Wick, Calla Rose Ostrander, David Bronner, the staff at Café Gratitude, and so many others I'm forgetting.

I owe a special thanks to the entire staff at Big Picture Ranch as well as the crew of *Kiss the Ground*, including my cinematographer Simon Balderas, who left his family for months on end to help lens this story and turn it into the visual feast that is *Kiss the Ground*. Thank you to Alexa Coughlin, our most dedicated employee, to Sam Gall, who has thrown his soul into our projects, and to our film editors: Sean Keenan, Anthony Ellison, Ryan Nichols, and Moment Dangquiing Lu. Thank you also to our executive producers for supporting this project, including Bill and Laurie Benenson, Anna Getty, Michelle Lerach, John Paul DeJoria, Craig McCaw, and Jena King.

Finally, thank you to the entire staff and volunteer network of the Kiss the Ground organization, especially Finian Makepeace, Lauren Tucker, and Karen Rodriguez. It is through your dedication to this cause, as well as the hard work of the numerous excellent NGOs working tirelessly on this issue, that the regenerative revolution is finally taking flight.

NOTES

INTRODUCTION

1. Alison Abbott, "Scientists Bust Myth That Our Bodies Have More Bacteria Than Human Cells," *Scientific American*, January 11, 2016, https://www.scientificamerican.com/article/scientists-bust-myth-that-our-bodies-have-more-bacteria-than-human-cells/.
2. The "Parable of the Soils" is also known as the "Parable of the Sower."

CHAPTER 1: SHOWDOWN IN PARIS

1. James Fallows, "Your Labor Day Syria Reader, Part 2: William Polk," *The Atlantic*, September 2, 2013, http://www.theatlantic.com/international/archive/2013/09/your-labor-day-syria-reader-part-2-william-polk/279255/.
2. Charles D. Keeling, "The Concentration and Isotopic Abundances of Carbon Dioxide in the Atmosphere," *Tellus* 12, no. 2 (1960): 200–203.
3. Prior to the creation of the character of James Bond, French writer Jean Bruce wrote a novel about a fictional secret agent named Hubert Bonisseur de La Bath, aka "OSS 117." Two hundred fifty OSS novels later, OSS 117 remains in French lore the "original" secret agent.
4. "Carbon Maths," Climate Consent, 2012, http://www.climateconsent.org/pages/carbonmaths.html.
5. Zhang-ting Huang, Yong-fu Li, Pei-kun Jiang, Scott X. Chang, Zhao-liang Song, Juan Liu, and Guo-mo Zhou, "Long-term Intensive Management Increased Carbon Occluded in Phytolith (PhytOC) in Bamboo Forest Soils," *Scientific Reports* 4 (January 2014), http://dx.doi.org/10.1038/srep03602.
6. 4 Pour 1000, "Understand the '4 per 1000' Initiative," 2015, http://4p1000.org/understand.
7. Tom Vilsack served as US secretary of agriculture from 2009 until 2017. He quit a week before the inauguration of Donald Trump.
8. "Lancement Du 4 Pour 1000: Stéphane Le Foll Salue La Mobilisation Internationale, plus de 100 États et Organisations Soutiennent L'initiative," *Alim'agri*, December

1, 2015, http://agriculture.gouv.fr/4-pour-1000-plus-de-100-etats-et-organisations-soutiennent-linitiative.

9. Speech is abbreviated. Full speech is located here: "COP 21: Le 1er Décembre en Images," *Alim'agri*, December 1, 2015, http://agriculture.gouv.fr/cop-21-le-1er-decembre-en-images.
10. Lord Deben and Lord Krebs, "The Good, the Bad, and the Ugly of the Paris Agreement," Committee on Climate Change, December 21, 2015, https://www.theccc.org.uk/2015/12/21/the-good-the-bad-and-the-ugly-of-the-paris-agreement/.
11. Henrik Selin and Adil Najam, "Paris Agreement on Climate Change: The Good, the Bad, and the Ugly," *The Conversation*, December 14, 2015, http://theconversation.com/paris-agreement-on-climate-change-the-good-the-bad-and-the-ugly-52242.
12. Deben and Krebs, "The Good, the Bad, and the Ugly."
13. Charles C. Mann, "Solar or Coal? The Energy India Picks May Decide Earth's Fate," *Wired*, December 2015, http://www.wired.com/2015/11/climate-change-in-india/.
14. Coral Davenport, "The Marshall Islands Are Disappearing," *New York Times*, December 1, 2015, http://www.nytimes.com/interactive/2015/12/02/world/The-Marshall-Islands-Are-Disappearing.html?_r=0.

CHAPTER 2: NAZIS AND NITROGEN

1. Diarmuid Jeffreys, *Hell's Cartel: IG Farben and the Making of Hitler's War Machine* (New York: Henry Holt & Co., 2010).
2. Oswald W. Knauth, "Farmers' Income," in *Income in the United States, Its Amount and Distribution, 1909–1919*, Volume II: *Detailed Report* (Washington, DC: National Bureau of Economic Research, 1922), 298–313, http://www.nber.org/chapters/c9420.pdf.
3. Elizabeth A. Ramey, *Class, Gender, and the American Family Farm in the 20th Century* (New York: Routledge, 2014).
4. John H. Perkins, *Insects, Experts, and the Insecticide Crisis: The Quest for New Pest Management Strategies* (New York: Plenum, 1982).
5. David Tilman, Kenneth G. Cassman, Pamela A. Matson, Rosamond Naylor, and Stephen Polasky, "Agricultural Sustainability and Intensive Production Practices," *Nature* 418 (August 2002): 671–77.
6. Thomson Gale, "Paul Hermann Müller Biography," *BookRags*, 2005.
7. "Ciba-Geigy Ltd. History," Funding Universe, http://www.fundinguniverse.com/company-histories/ciba-geigy-ltd-history/.
8. "DDT General Fact Sheet," *SpringerReference: National Pesticide Information Center*, Oregon State University and the US Environmental Protection Agency, December 1999.

9. Ibid.
10. Daniel Smith, "Worldwide Trends in DDT Levels in Human Breast Milk," *International Journal of Epidemiology* 28, no. 2 (1999): 179–88.
11. Lindsey Konkel, "DDT Linked to Fourfold Increase in Breast Cancer Risk," *National Geographic*, June 16, 2015, http://news.nationalgeographic.com/2015/06/15616-breast-cancer-ddt-pesticide-environment/.
12. Dr. P. Toft, "2,4-D in Drinking-water," World Health Organization, 2003, http://www.who.int/water_sanitation_health/dwq/chemicals/24D.pdf.
13. Ibid.
14. Caroline Cox, "Herbicide Factsheet: 2,4-D," *Journal of Pesticide Reform* 25, no. 4 (2005): 10–15.
15. Ibid.
16. Enlist Duo Herbicide, *Enlist Duo Herbicide*, http://www.enlist.com/en/how-it-works/enlist-duo-herbicide.
17. "Slack Science Destroys Monsanto Breast Milk Study," *Sustainable Pulse*, July 27, 2015, http://sustainablepulse.com/2015/07/27/slack-science-destroys-monsanto-breast-milk-study/#.V2Muclc5P9o.
18. S. Thongprakaisang, A. Thiantanawat, N. Rangkadilok, T. Suriyo, and J. Satayavivad, "Glyphosate Induces Human Breast Cancer Cells Growth via Estrogen Receptors," *Food and Chemical Toxicology* (September 2013), http://www.ncbi.nlm.nih.gov/pubmed/23756170.
19. Gilles-Eric Séralini, Emilie Clair, Robin Mesnage, Steeve Gress, Nicolas Defarge, Manuela Malatesta, Didier Hennequin, and Joël Spiroux de Vendômois, "Republished Study: Long-term Toxicity of a Roundup Herbicide and a Roundup-tolerant Genetically Modified Maize," *Environmental Sciences Europe* (June 2014): 1–17, https://enveurope.springeropen.com/articles/10.1186/s12302-014-0014-5.
20. Tamsyn M. Uren Webster and Eduarda M. Santos, "Global Transcriptomic Profiling Demonstrates Induction of Oxidative Stress and of Compensatory Cellular Stress Responses in Brown Trout Exposed to Glyphosate and Roundup," *BMC Genomics* (January 2015): 1–14, https://bmcgenomics.biomedcentral.com/articles/10.1186/s12864-015-1254-5.
21. Jason Best, "Monsanto Weed Killer Is Discovered in Major Cereal Brands," *TakePart*, April 19, 2016.
22. M. S. Majewski, R. H. Coupe, W. T. Foreman, and P. D. Capel, "Pesticides in Mississippi Air and Rain: A Comparison between 1995 and 2007," *Environmental Toxicology and Chemistry* 33, no. 6: 1283–93, doi:10.1002/etc.2550.
23. Shireen, "Another Strike Against GMOs—The Creation of Superbugs and Superweeds," GMO Inside, March 31, 2014, http://gmoinside.org/another-strike-gmos-creation-superbugs-superweeds/.
24. Curtis Mowry, Adam Pimentel, Elizabeth Sparks, and Brittany Hanlon, "Occurrence

and Distribution in Streams and Ground Water," in Robert J. Gilliom, Jack E. Barbash, and Charles G. Crawford, *The Quality of Our Nation's Waters: Pesticides in the Nation's Streams and Ground Water, 1992–2001* (Reston, VA: US Geological Survey, 2006), 41–66, http://pubs.usgs.gov/circ/2005/1291/pdf/circ1291_chapter4.pdf.

25. "Atrazine: Chemical Summary," *Toxicity and Exposure Assessment for Children's Health* (Washington, DC: Environmental Protection Agency, 2007), 1–12, https://archive.epa.gov/region5/teach/web/pdf/atrazine_summary.pdf.
26. Tyrone B. Hayes, Lloyd L. Anderson, Val R. Beasley, Shane R. de Solla, Taisen Iguchi, et al., "Demasculinization and Feminization of Male Gonads by Atrazine: Consistent Effects across Vertebrate Classes," *Journal of Steroid Biochemistry and Molecular Biology* 127, nos. 1–2: 64–73, doi:10.1016/j.jsbmb.2011.03.015.
27. Sheila Kaplan, "Studies Show Pesticides Harming Salinas Valley Children," *Investigative Reporting Workshop*, December 21, 2010, http://investigativereportingworkshop.org/investigations/toxic-influence/story/studies-show-pesticides-harming-children.
28. Janie F. Shelton, Estella M. Geraghty, Daniel J. Tancredi, Lora D. Delwiche, Rebecca J. Schmidt, Beate Ritz, Robin L. Hansen, and Irva Hertz-Picciotto, "Neurodevelopmental Disorders and Prenatal Residential Proximity to Agricultural Pesticides: The CHARGE Study," *Children's Health* 122, no. 10 (2014): 1103–110, http://ehp.niehs.nih.gov/wp-content/uploads/122/10/ehp.1307044.alt.pdf.
29. Jim Feuer, "Study Links Exposure to Common Pesticide With ADHD in Boys," Cincinnati Children's, June 1, 2015, http://www.cincinnatichildrens.org/news/release/2015/study-links-pesticide-ADHD-in-boys-06-01-2015/.
30. Virginia A. Rauh, Robin Garfinkel, Frederica P. Perera, Howard F. Andrews, Lori Hoepner, Dana B. Barr, Ralph Whitehead, Deliang Tang, and Robin W. Whyatt, "Impact of Prenatal Chlorpyrifos Exposure on Neurodevelopment in the First 3 Years of Life Among Inner-City Children," *Pediatrics* 118, no. 6 (2006): E1845–1859.
31. Sean Poulter, "Up to 98% of Our Fresh Food Carries Pesticides: Proportion of Produce with Residues Doubles in a Decade," *Daily Mail*, August 28, 2013, http://www.dailymail.co.uk/news/article-2405078/Up-98-fresh-food-carries-pesticides-Proportion-produce-residues-doubles-decade.html.
32. Environmental Working Group, "EWG's Shopper's Guide to Pesticides in Produce," https://www.ewg.org/foodnews/summary.php.
33. Peter Wood, "USDA Releases 2014 Annual Summary for Pesticide Data Program: Report Confirms that Pesticide Residues Do Not Pose a Safety Concern for U.S. Food," US Department of Agriculture: Agriculture Marketing Service, January 11, 2016, https://www.ams.usda.gov/press-release/usda-releases-2014-annual-summary-pesticide-data-program-report-confirms-pesticide.
34. *Pesticide Data Program: Annual Summary (2014)*, USDA: Agriculture Marketing Service, January 2016, https://www.ams.usda.gov/sites/default/files/media/2014%20PDP%20Annual%20Summary.pdf.

35. "Food and Agriculture," data files, Earth Policy Institute, http://www.earth-policy.org/data_center/C24.
36. Ibid. and FAOSTAT, http://www.fao.org/faostat/en.
37. Wenonah Hauter, *Foodopoly: The Battle over the Future of Food and Farming in America* (New York: New Press, 2012).
38. Gary Schnitkey, "Cost Cutting for 2016: Budgeting for $4 Corn and $9.25 Soybeans," *Farmdoc Daily*, University of Illinois Department of Agricultural and Consumer Economics, August 4, 2015, http://farmdocdaily.illinois.edu/2015/08/cost-cutting-for-2016-budgeting-for-corn-soybeans.html.
39. Action Group on Erosion, Technology, and Concentration (ETC Group), "Seeds & Genetic Diversity," http://www.etcgroup.org/issues/seeds-genetic-diversity.
40. Donald R. Davis, "Declining Fruit and Vegetable Nutrient Composition: What Is the Evidence?" *HortScience* 44, no. 1 (2009): 15–19, http://hortsci.ashspublications.org/content/44/1/15.full.pdf+html.
41. Duke University, "How Much Water Does U.S. Fracking Really Use? Water Used in Fracking Makes up Less than One Percent of Total Industrial Water Use Nationwide, Study Finds," *ScienceDaily*, September 15, 2015, www.sciencedaily.com/releases/2015/09/150915135827.htm.
42. Natasha Geiling, "California Farmers Are Watering Their Crops with Oil Wastewater, and No One Knows What's In It," *Think Progress*, May 5, 2015, http://thinkprogress.org/climate/2015/05/05/3654388/california-drought-oil-waste-water-agriculture/.
43. Alex Nussbaum and David Wethe, "California Farms Are Using Drilling Wastewater to Grow Crops," *Bloomberg*, July 8, 2015, http://www.bloomberg.com/news/articles/2015-07-08/in-california-big-oil-finds-water-is-its-most-prized-commodity.
44. Martha Rosenberg and Ronnie Cummins, "Monsanto's Evil Twin: Disturbing Facts About the Fertilizer Industry," Organic Consumers Association, April 5, 2016.

CHAPTER 3: ENDLESS SUMMER

1. "California Economy," Netstate, http://www.netstate.com/economy/ca_economy.htm.
2. California Department of Food and Agriculture, "California Agricultural Production Statistics," https://www.cdfa.ca.gov/statistics/.
3. California Avocado Commission, "Irrigating Avocados Fact Sheet," http://www.californiaavocadogrowers.com/sites/default/files/documents/Irrigating-Avocados-Fact-Sheet.pdf.
4. Two hundred avocadoes is around one hundred pounds of fruit. Looking at the water-to-fruit ratio, if we follow the instructions for growing avocadoes, we need to use 2.75 million gallons of water to produce thirty thousand pounds of fruit. Even if our trees were better than perfect—ideal, perhaps—and they doubled

their output to two hundred pounds of fruit each, they would grow only sixty thousand pounds of avocadoes.

5. Ian James, "USGS Estimates Vast Amounts of Water Used in California," *The Desert Sun*, August 21, 2014, http://www.desertsun.com/story/news/environment/2014/08/21/usgs-estimates-vast-amounts-water-used-california/14400333/.
6. Dana Gunders, "Wasted: How America Is Losing Up to 40 Percent of Its Food from Farm to Fork to Landfill," *National Resources Defense Council Issue Paper* B 12, no. 6 (August 2012): 4–21, https://www.nrdc.org/sites/default/files/wasted-food-IP.pdf.
7. Daniel Zohary, Maria Hopf, and Ehud Weiss, *Domestication of Plants in the Old World: The Origin and Spread of Domesticated Plants in Southwest Asia, Europe, and the Mediterranean Basin* (Oxford, UK: Oxford University Press, 2012).
8. Steven J. Mithen, *After the Ice: A Global Human History, 20,000–5000 BC* (Cambridge, MA: Harvard University Press, 2004).
9. Jacques Cauvin and Trevor Watkins, *The Birth of the Gods and the Origins of Agriculture* (Cambridge, UK: Cambridge University Press, 2000).
10. As of May 2017, the PDF was available here: http://epsc413.wustl.edu/Lowdermilk_Conquest_USDA.pdf.
11. Walter Clay Lowdermilk, *Conquest of the Land through Seven Thousand Years, by W. C. Lowdermilk* (Washington, DC: US Government Printing Office, 1953).
12. H. E. Dregne, "Desertification of Arid Lands," in F. El-Baz and M. H. A. Hassan, eds., *Physics of Desertification* (Dordrecht, Netherlands: Martinus, Nijhoff, 1986).
13. Helmut J. Geist, *The Causes and Progression of Desertification* (Aldershot, UK: Ashgate Publishing, 2005).
14. R. W. A. Hutjes, P. Kabat, S. W. Running, W. J. Shuttleworth, C. Field, B. Bass, M. A. F. Da Silva Dias, R. Avissar, A. Becker, M. Claussen, A. J. Dolman, R. A. Feddes, M. Fosberg, Y. Fukushima, J. H. C. Gash, L. Guenni, H. Hoff, P. G. Jarvis, et al., "Biospheric Aspects of the Hydrological Cycle," *Journal of Hydrology* 212–213 (1998): 1–21.
15. W. Ripl, "Mathematical Modelling in Limnology Management of Water Cycle and Energy Flow for Ecosystem Control: The Energy-transport-reaction (ETR) Model," *Ecological Modelling* 78, no. 1 (1995): 61–76.
16. R. A. Pielke Sr., "Influence of the Spatial Distribution of Vegetation and Soils on the Prediction of Cumulus Convective Rainfall," *Reviews of Geophysics* 39, no. 2: 151–177, doi:10.1029/1999RG000072.
17. Environmental Protection Agency, "Heat Island Effect," https://www.epa.gov/heat-islands.
18. Sanden Totten, "LA Area Has Highest Urban Heat Island Effect in California," 89.3 KPCC, Southern California Public Radio, September 21, 2015, http://www.scpr.org/news/2015/09/21/54511/la-area-has-highest-urban-heat-island-effect-in-ca/.

19. Pandi Zdruli, Marcello Pagliai, Selim Kapur, and Angel Faz Cano, eds., *Land Degradation and Desertification: Assessment, Mitigation and Remediation* (Dordrecht, Netherlands: Springer, 2010).
20. Dregne, "Desertification of Arid Lands."
21. Alex Park and Julia Lurie, "It Takes How Much Water to Grow an Almond?!" *Mother Jones*, February 24, 2014, http://www.motherjones.com/environment/2014/02/wheres-californias-water-going.
22. Todd C. Frankel, "New NASA Data Show How the World Is Running Out of Water," *Washington Post*, June 16, 2015, https://www.washingtonpost.com/news/wonk/wp/2015/06/16/new-nasa-studies-show-how-the-world-is-running-out-of-water.
23. Zafar Adeel, Janos Bogardi, et al., *Overcoming One of the Greatest Environmental Challenges of Our Times: Re-thinking Policies to Cope with Desertification* (Hamilton, Ontario: United Nations University Institute for Water, Environment, and Health, 2007), http://inweh.unu.edu/wp-content/uploads/2013/05/Re-thinkingPolicietoCopewithDesertification.pdf.
24. John Wendle, "The Ominous Story of Syria's Climate Refugees," *Scientific American*, December 17, 2015, http://www.scientificamerican.com/article/ominous-story-of-syria-climate-refugees/.
25. UNHCR: The UN Refugee Agency, "Figures at a Glance," http://www.unhcr.org/en-us/figures-at-a-glance.html.
26. UNHCR: The UN Refugee Agency, *Global Trends: Forced Displacement in 2015*, http://www.unhcr.org/576408cd7.pdf.

CHAPTER 4: MEET THE REGENETARIANS

1. Broad and Blue, "Revenue Earned at Magic Kingdom since Page Load," per May 22, 2017, http://www.broadbandblue.com/disney-revenue.
2. Many thanks to the Kerr Center for Sustainable Agriculture and George Kuepper for their report *A Brief Overview of the History and Philosophy of Organic Agriculture* (2010), http://kerrcenter.com/wp-content/uploads/2014/08/organic-philosophy-report.pdf.
3. David McCandless, "How Many Gigatons of CO2?" Information Is Beautiful, February 2016, http://www.informationisbeautiful.net/visualizations/how-many-gigatons-of-co2/. Approximately 50 percent of that CO_2 has since been absorbed by oceans and land.

CHAPTER 5: THE BUFFALO BANK ACCOUNT

1. Robb Campbell, "S. D. Native Grass Map," US Soil Conservation Service, 1997, http://www.augie.edu/dept/biology/tieszen/rcampbell/SCS1942.gif.
2. Bureau of Sport Fisheries and Wildlife, "The American Buffalo," *Conservation Note* 12 (January 1965).

3. Philip St. George Cooke, "Scenes in the West; or, A Night on the Santa Fe Trail, No. III," *Southern Literary Messenger* (February 1842).
4. Horace Greeley, *An Overland Journey, from New York to San Francisco in the Summer of 1859* (New York: C. M. Saxton, Barker & Co., 1860).
5. Shepard Krech III, "Buffalo Tales: The Near-Extermination of the American Bison," Brown University National Humanities Center, http://nationalhumanitiescenter.org/tserve/nattrans/ntecoindian/essays/buffalob.htm.
6. Joshua Lederberg, Robert E. Shope, and Stanley Oaks Jr., eds., *Emerging Infections: Microbial Threats to Health in the United States* (Washington, DC: National Academies Press, 1992), http://www.nap.edu/read/2008/chapter/3#24.
7. Bureau of Sport Fisheries and Wildlife, "The American Buffalo."
8. J. Knox Jones Jr., "Roe, Frank Gilbert. *The North American Buffalo: A Critical Study of theSpecies in Its Wild State*," *Journal of Mammalogy* 52, no. 2: 487.
9. Bureau of Sport Fisheries and Wildlife, "The American Buffalo."
10. It takes thirty-three cubic yards of compost to cover an acre of soil to a quarter inch deep. With about 850,000 residents, San Francisco produces 488 cubic yards of compost per day. The United States has about 320 million people (thus the country is about 376 times bigger than San Francisco). 376 x 488 cubic yards = 183,488 cubic yards of possible compost production for the nation per day. Divide by 33, and that equals 5,560 acres that can be covered each day by the total amount of theoretical compost that can be produced in the United States. The United States has approximately 433,343,765 arable acres. Producing compost 365 days a year from 100 percent of the US population could cover about 2,029,400 acres—in other words, about 0.5 percent of the total cropland soils of the entire country could be covered if we converted every available food scrap into compost. Looking at total agricultural and rangeland (approximately 915 million acres), city-derived compost could cover less than 0.25 percent.
11. The biosolids number may be a stretch, as much of the "disinfected" sewage waste produced in the United States is already used on agriculture fields. See *Biosolids Technology Fact Sheet: Land Application of Biosolids*, Volume 832, Edition F, Series 064 (Washington, DC: US Environmental Protection Agency, Office of Water, September 2000), https://www3.epa.gov/npdes/pubs/land_application.pdf.
12. I used 1.33 as the multiplier to convert tons to cubic yards of finished compost.
13. USDA Natural Resources Conservation Service, "Animal Manure Management: RCA Issue Brief #7," December 1995, http://www.nrcs.usda.gov/wps/portal/nrcs/detail/national/technical/nra/dma/?cid=nrcs143_014211.
14. "Lexicon Terms: Ecological Memory, Ecosystem Restoration and Novel Ecosystems," Framing a Modern Mess, January 15, 2011, https://cityarchpruittigoe.

wordpress.com/2011/01/15/lexicon-term-ecological-memory-ecosystem-restoration-and-novel-ecosystems/.

CHAPTER 6: HOME ON THE RANGE

1. National Chicken Council, "Per Capita Consumption of Poultry and Livestock, 1965 to Estimated 2016, in Pounds," September 21, 2016, http://www.nationalchickencouncil.org/about-the-industry/statistics/per-capita-consumption-of-poultry-and-livestock-1965-to-estimated-2012-in-pounds/.
2. Check out their website (http://wildernessawareness.org) and their Facebook page (https://www.facebook.com/wildernessawareness/).
3. Frances Dinkelspiel, "What Did California Look Like Before People?" *Berkeleyside*, November 9, 2010, http://www.berkeleyside.com/2010/11/09/what-did-california-look-like-before-people/.
4. P. J. Gerber, H. Steinfeld, B. Henderson, A. Mottet, C. Opio, J. Dijkman, A. Falcucci, and G. Tempio, *Tackling Climate Change Through Livestock—A Global Assessment of Emissions and Mitigation Opportunities*, Food and Agriculture Organization of the United Nations, Rome, 2013.
5. Environmental Protection Agency, "Overview of Greenhouse Gases," February 14, 2017, http://www3.epa.gov/climatechange/ghgemissions/gases/co2.html.
6. "Grass-Fed Beef Has Bigger Carbon Footprint: Discovery News," *Seeker*, February 11, 2013, http://news.discovery.com/earth/grass-fed-beef-grain.htm.
7. Alexander Hristov, "Wild Ruminants Burp Methane, Too (Dairy)," *PennState Extension*, April 21, 2011, http://extension.psu.edu/animals/dairy/news/2011/wild-ruminants-burp-methane-too.
8. Fred Pearce, "What Is Causing the Recent Rise in Methane Emissions?" *Yale Environment 360*, October 25, 2016, http://e360.yale.edu/features/methane_riddle_what_is_causing_the_rise_in_emissions.

CHAPTER 7: THE SOILVANGELIST

1. David Pimentel and Mario Giampietro, *Food, Land, Population and the U.S. Economy* (Washington, DC: Carrying Capacity Network, 1994).
2. "The Five Main 'Crop Belts' of American Agriculture," *The American Farmer*, 1932, http://imgur.com/r3s1UxY.
3. David J. Wishart, ed., *Encyclopedia of the Great Plains* (Lincoln: University of Nebraska, 2011), http://plainshumanities.unl.edu/encyclopedia/.
4. "Soil Conservation in the New Deal Congress," *US House of Representatives: History, Art & Archives* (Library of Congress, 1935), http://history.house.gov/Historical-Highlights/1901-1950/Soil-Conservation-in-the-New-Deal-Congress/.
5. USDA Economic Research Service, "Data Files: U.S. and State-Level Farm Income and Wealth Statistics," http://www.ers.usda.gov/data-products/farm-

income-and-wealth-statistics/data-files-us-and-state-level-farm-income-and-wealth-statistics.aspx.

CHAPTER 8: BISMARCK OR BUST

1. "The Basics of Crop Insurance," Proag.com, http://www.proag.com/basics-of-crop-insurance.
2. Bionutrient Food Association, "Brix," http://bionutrient.org/bionutrient-rich-food/brix.
3. US Department of Agriculture, "2012 Census Highlights," May 2014, https://www.agcensus.usda.gov/Publications/2012/Online_Resources/Highlights/Farm_Demographics/.

CHAPTER 9: A NEW PLATE

1. Lisa Abend, "The Cult of the Celebrity Chef Goes Global," *Time*, June 21, 2010, http://content.time.com/time/magazine/article/0,9171,1995844-3,00.html.
2. Roberto A. Ferdman, "The Slow Death of the Home-cooked Meal," *Washington Post*, March 5, 2015, https://www.washingtonpost.com/news/wonk/wp/2015/03/05/the-slow-death-of-the-home-cooked-meal/.
3. "Industrial Agriculture and Small-scale Farming," Global Agriculture: Agriculture at a Crossroads, http://www.globalagriculture.org/report-topics/industrial-agriculture-and-small-scale-farming.html.
4. Ibid.
5. Ayman Sawaf, *Sacred Commerce: The Rise of the Global Citizen* (Ojai, CA: Sacred Commerce, 2007).

CHAPTER 10: THE REGENERATIVE REVOLUTION

1. The USDA has a colorful webpage with all their various loan programs listed below (clicking deeper into the pages will get you to the PDF loan applications themselves): http://www.fsa.usda.gov/programs-and-services/farm-loan-programs/index.